Andrew H Baker

Short and comprehensive course of geometry and trigonometry

designed for general use for general use in schools and colleges

Andrew H Baker

Short and comprehensive course of geometry and trigonometry
designed for general use for general use in schools and colleges

ISBN/EAN: 9783741192654

Manufactured in Europe, USA, Canada, Australia, Japa

Cover: Foto ©Thomas Meinert / pixelio.de

Manufactured and distributed by brebook publishing software
(www.brebook.com)

Andrew H Baker

Short and comprehensive course of geometry and trigonometry

A

SHORT AND COMPREHENSIVE COURSE

OF

GEOMETRY AND TRIGONOMETRY;

DESIGNED FOR GENERAL USE IN

SCHOOLS AND COLLEGES.

BY

ANDREW H. BAKER, A.M., Ph.D.

NEW YORK:

P. O'SHEA, PUBLISHER,

37 BARCLAY STREET.

1878.

PREFACE.

GEOMETRY, like every other science, has but few principles, which, if systematically arranged and thoroughly developed, may be readily comprehended, and indelibly impressed upon the mind.

Plane Geometry may be said to begin and end with the circle. The angles formed at the center by the radii should be treated of first; next, inscribed angles, and an inscribed triangle; then, a hexagon and an equilateral triangle; inscribed and circumscribed squares; regular and irregular polygons; and finally return to the circle. In the demonstrations of the propositions derived from these figures, every principle of Plane Geometry is developed; and Solid Geometry has no distinct principles.

In teaching Solid Geometry, I much prefer the use of blocks. The beginner, at least, will be greatly benefited by having a material object to inspect and compute, until he becomes thoroughly acquainted with all its properties; after which, he may employ his imagination as he likes, and conceive figures of every shape and form. The system of object teaching favors this method.

In preparing this treatise, I have aimed especially at simplicity and brevity. The former, that it may be within the grasp of every student; the latter, that the memory may not be overburdened, and too much time occupied in acquiring a thorough knowledge of the science.

Although I have stated that beginners derive benefits from material figures, I do not thereby wish to intimate that Geometry presents no opportunity for the exercise of the imagination; whilst, in truth, no other science presents so wide a field for the exercise of this faculty. We pierce the most distant points of the celestial concave with straight lines, and with arcs of great circles; measure and compute the distance and size of the farthest stars, thereby rendering what appeared imaginary, matters of fact.

Simplicity and brevity are not only important, but they are absolutely necessary, in order that mankind generally may acquire a thorough knowledge of pure mathematics; after which, any branch of applied mathematics may be pursued with ease and advantage.

The design of the Trigonometry is the same as of the Geometry.

With these impressions, I dedicate this volume to the American Youth; and if it prove that I have plucked a few thorns from the rugged path of science, and strewn a few flowers therein, I shall not regret the arduous task.

AUTHOR.

ELEMENTS OF GEOMETRY.

BOOK I.

DEFINITIONS.

1. Elementary Geometry treats of the properties, relations, and measurement of magnitudes.

2. Magnitudes have one, two, or three dimensions; as,
A line has only one dimension, viz., length.
A surface has two, length and breadth.
And a solid has three, length, breadth, and thickness.

3. Plane Geometry takes its name from the plane, as each figure is upon one plane.

REM.—As every figure, and every part of it, is on the same plane, it is not necessary to repeat "on the same plane."

4. A Mathematical Plane is a surface of indefinite extent, such that if a straight edge or rule be applied to it, the edge or rule will coincide with it, in every position.

5. Lines are of two classes, *straight* and *curved*.

A Straight Line has everywhere the same direction, or it may be said to have two directions, exactly opposite each other, from any point in the line.

A Curved Line, or simply a **Curve,** constantly changes its direction.

6. Surfaces are of two classes, *plane* and *curved*.

A Plane Surface corresponds to the mathematical plane, or a portion of it, and it may have any position whatever; that is, it may be horizontal or vertical, or it may be oblique.

A Curved Surface is such that if a straight rule be applied to it, the rule will not coincide with it in every position; as the surface of a sphere or of a cylinder.

7. A **Point** has position only; as, any particular place in a line or plane, and the extremities of lines, are called points.

8. A **Circle** is a portion of a plane bounded by a curved line, every point of which is equally distant from a point within called the center.

9. The curved line is called the **Circumference,** and any part of it an **Arc.**

10. A **Polygon** is a portion of a plane bounded by straight lines called *sides.*

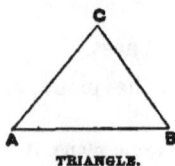

A polygon of three sides is called a **Triangle.**
A polygon of four sides is called a **Quadrilateral.**
A polygon of five sides is called a **Pentagon.**

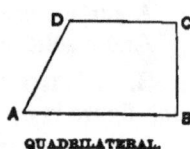

TRIANGLE.

QUADRILATERAL.

A polygon of six sides is called a **Hexagon,** etc.

11. The divergence of any two sides from their point of intersection is called an **Angle** of the polygon ; and the number of angles will always be the same as the number of sides of the polygon.

12. A triangle having two equal sides is called an **Isosceles** triangle.
A triangle having three equal sides is called an **Equilateral** triangle.
A triangle having all its sides unequal is called a **Scalene** triangle.

13. Two lines are **Parallel** when they are everywhere equally distant, and hence will never meet.

14. A quadrilateral having its opposite sides respectively parallel is called a **Parallelogram.**

15. A quadrilateral having only two sides parallel is called a **Trapezoid.**

16. A **Regular Polygon** has all its sides and angles respectively equal.

17. A regular quadrilateral is termed a **Square.**

18. A circle, a polygon, etc., are termed **Geometrical Figures.**

19. An **Angle** may be designated by the letter at its vertex, or by three letters, the letter at the vertex occupying the middle place, and the letters at the extremities of its sides holding the first and last places; thus, the angle A in the triangle ABC is designated angle BAC.

20. When the angles of a quadrilateral are right angles, and the opposite sides respectively equal, it is termed a **Rectangle**.

21. The circumference of a circle is divided into 360 equal parts, called **Degrees**, and if radii be drawn to each point marking the degrees, there will be 360 angles, each of one degree.

GEOMETRICAL TERMS.

1. An **Axiom** is a self-evident truth.

2. A **Theorem** is a truth which requires a demonstration.

3. A **Problem** is a question which requires a solution.

4. Axioms, Theorems, and Problems are **Propositions**.

5. A **Corollary** is an obvious consequence of one or more propositions, or of a definition.

6. A **Scholium** is a remark upon something which precedes.

GENERAL AXIOMS.

1. Magnitudes which are equal to the same magnitude are equal to each other.

2. If equals be added to equals, the sums will be equal.

3. If equals be subtracted from equals, the remainders will be equal.

4. If equals be added to unequals, the sums will be unequal.

5. If equals be subtracted from unequals, the remainders will be unequal.

6. If equals be multiplied by equals, the products will be equal.

7. If equals be divided by equals, the quotients will be equal.

8. The whole is greater than any of its parts.

9. The whole is equal to the sum of all its parts.

10. Like powers and like roots of equals are equal.

SPECIAL AXIOMS.

1. A straight line is the shortest distance between two points.

2. Between two points only one straight line can be drawn.

3. Two fixed points through which a line passes determine its direction.

Cor. 1.—Two straight lines, having two points common, form one and the same straight line.

Cor. 2.—Two straight lines intersect at but one point.

4. Two straight lines, starting from the same point and taking the same direction, form one and the same straight line.

5. Two straight lines, starting from the same point and taking different directions, form an angle.

6. Two straight lines starting from different points and taking the same direction, are parallel.

7. If two lines are each parallel to a third, they will be parallel to each other.

8. Only one perpendicular can be drawn to a straight line, either from a point without the line, or from a point on the line.

Describe a circle, and show the relation of its properties, and also that of a straight line touching it at one point.

Take a string of any definite length, say six inches, attach a pin to one end, and a chalk point to the other end. Fasten the pin at any point in a plane as a center, and, with the string at

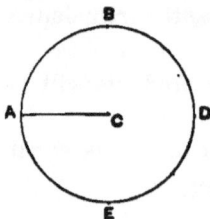

full stretch, revolve the chalk point around the center, until it reaches the point from which it started; thus, let CA represent the string, C the center-pin, and A the chalk-point. ABDE is the chalk-line made by the revolution of CA; every point in the line ABDE will be at the distance CA from the center; the curved line ABDE is the circumference of the circle; the portion of the plane enclosed by it is the circle; and CA is the radius.

DEF. 1.—Any straight line, as AB, passing through the center and terminating on the circumference, is a **Diameter.**

COR.—A diameter is twice the radius.

DEF. 2.—Any straight line as DE, touching at but one point as F, is a **Tangent** to the circumference.

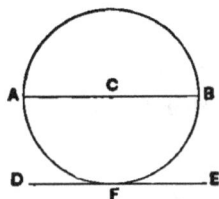

REM.—A circumference can be described with a pair of dividers; the distance between the points is the **Radius** of the circle.

Two radii drawn from the center of a circle to its circumference, form an angle of as many degrees as is contained in the arc intercepted by its sides.

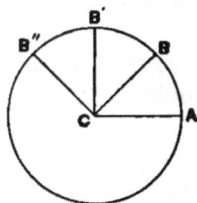

DEF. 1.—When the angle is 90 degrees, it is called a **Right Angle.**

DEF. 2.—When the angle is less than 90 degrees, it is called an **Acute Angle.**

DEF. 3.—When the angle is greater than 90 degrees, it is called an **Obtuse Angle.**

DEF. 4.—The **Complement** of an angle is the difference between the angle and 90 degrees.

COR.—If the sum of two angles is 90 degrees, the one is the complement of the other.

DEF. 5.—The **Supplement** of an angle is the difference between the angle and 180 degrees.

COR.—If the sum of two angles is 180 degrees, the one is the supplement of the other.

THEOREM I.

The diameter of a circle bisects the circle and its circumference.

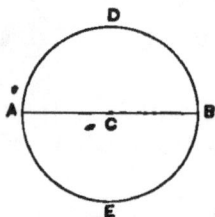

Let AB be the diameter of the circle ADBE.

Revolve the part ADB upon AB as an axis, until it falls upon AEB. The arc ADB will coincide with the arc AEB; otherwise some points in the circumference would be unequally distant from the center of the circle; hence the part of the circle ADB is equal to the part AEB; and the arc ADB is equal to the arc AEB.

Cor.—Each one of the two equal parts of the circle is a semi-circle, and the corresponding arcs are semi-circumferences.

THEOREM II.

An angle at the center of the circle is measured by the arc intercepted by its sides.

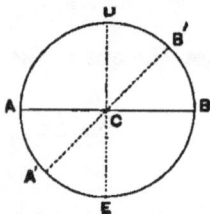

Let C be the center and AB the diameter of a circle, A'B' a two-pointed needle, with a pivot at the center C about which it revolves. Since A'B' passes through the center and terminates in the circumference, it is a diameter, and in every position bisects the circle and its circumference (Theorem 1); hence, as A' is moved towards E, B' moves towards D; the arcs AA' and BB' are constantly equal; and the radii CA and CA', also CB and CB', make equal angles at the center C.

When A' reaches E, 90 degrees from A; B' will be at D, 90 degrees from B; and there will be four equal angles at C, each 90 degrees; and the diameters are said to be at right angles, or perpendicular to each other.

As the arc AA' increases by one, two, etc., degrees, so also the angle ACA' increases by the same number of degrees.

THEOREM III.

If one straight line intersect another straight line, the sum of any two adjacent angles will be equal to two right angles.

Let the two straight lines intersect at C, then with C as a center and any radius describe a circumference cutting the lines at A, E, B and D. Since AB is a diameter, the two angles ACE and ECB will be measured by the sum of the two arcs AE and EB, which is equal to a semi-circumference or two right angles. So also the two angles ACD and BCD, and since DE is a diameter, the sum of ACD and ACE is equal to two right angles, also the sum of ECB and BCD.

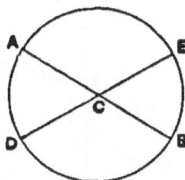

Cor. 1.—Vertical angles are equal, as each one is the supplement of the same angle; thus, ACD and ECD is each the supplement of ACE, or its equal BCD.

Cor. 2.—The sum of all the angles at a point on each side of a straight line is equal to two right angles ; and the sum of all the angles around a point is equal to four right angles.

Cor. 3.—Equal arcs have equal radii and are like parts of equal circumferences.

Cor. 4.—Equal angles have equal arcs, and equal arcs have equal chords, the radii being equal.

Scho.—If several circumferences, with different radii, be described from the same center, the circumferences will be parallel.

THEOREM IV.

The diameter of a circle is greater than any other chord.

Let AB be the diameter and BD a chord of a circle. Draw the radius CD, which is equal to CA; BD is less than the sum of BC and CD, (Special Axiom 1); but BC + CD = AC + CB = AB ; therefore DB is less than AB, or AB > DB.

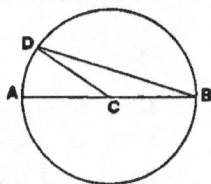

THEOREM V.

If two straight lines meet a third line, making any two angles which are similarly situated with regard to the two lines, and on the same side of the third line equal, then will the two lines be parallel. The converse is also true.

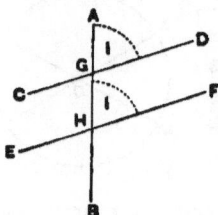

If the two lines CD and EF meet AB, making the angles AGD and AHF equal, then will CD and EF be parallel.

CD and EF may be regarded as starting at different points G and H, and as they make the angles AGD and AHF equal, they take the same direction and are therefore parallel. (Special Axiom 6.)

The converse is necessarily true.

COR. 1.—The same is true, when the equal angles are right angles ; hence, two lines perpendicular to a third are parallel.

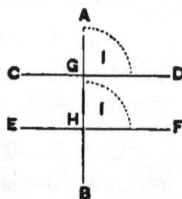

COR. 2.—Since the angles marked ı and ı are equal, and their vertical angles are also equal, hence four of these angles are equal; and as each of the remaining four is supplementary to one of these, they are consequently equal.

COR. 3.—If one of these angles is acute, four will be acute, and the other four will be obtuse ; but if one is a right angle, all will be right angles.

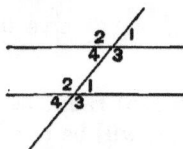

DEF. 1.—Angles similarly situated are called corresponding angles.

COR. 1.—If two parallels are cut by a third line, the corresponding angles will be equal.

COR. 2.—The interior angles on the same side are supplementary.

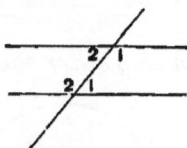
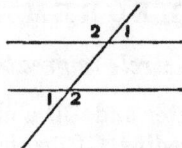
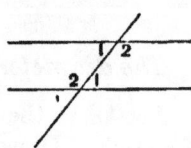

COR. 2.   COR. 3.   COR. 4.

COR. 3.—The alternate exterior angles are equal.

COR. 4.—The alternate interior angles are equal.

SCHO.—In the above figures the same numbers indicate pairs.

THEOREM VI.

Two angles, having their sides parallel and lying in the same or in opposite directions, are equal.

Let AB and DE be parallel, also BC and EF, and lying in the same direction; then will the angles ABC and DEF be equal.

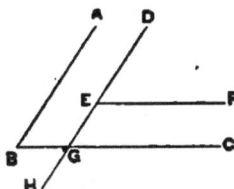

Produce DE to H, cutting BC in G. Since the parallels are cut by DH, the corresponding angles DEF and DGC are equal; and as BC cuts the parallels DH and AB, the angles DGC and ABC are corresponding angles and hence are equal; therefore, the angle ABC is equal to the angle DEF. (Ax. 1.)

The angles DGC and BGH are vertical angles, therefore equal. Consequently the angles ABC and BGH are equal.

THEOREM VII.

Two angles, having their sides respectively perpendicular, are equal or supplementary.

Let the sides of the angle EAD be respectively perpendicular to the sides of the angle BAC; and also the sides of EAD' perpendicular to the sides of BAC.

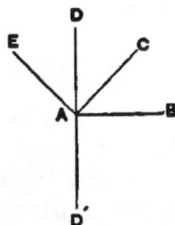

1st. The angles BAD and CAE are right angles; from each take the angle CAD, and there remains the angle BAC equal to the angle DAE.

2d. The angle EAD' is supplementary to EAD; so also of its equal BAC.

THEOREM VIII.

If, in a circle, two diameters be drawn at right angles, and several chords be drawn parallel to one of the diameters, and at the extremities of the other diameter, lines be drawn parallel to the chords,

1st. *The chords will be bisected by the perpendicular diameter.*

2d. *The lines at the extremities of the same diameter, will be tangents to the circumference.*

3d. *Any two parallels will intercept equal arcs of the circumference.*

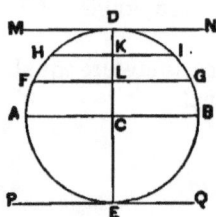

Since DE is a diameter, DAE is a semicircle, and if it be revolved upon DE as an axis, until it fall upon DBE, the two semicircles will coincide; and since all the angles made with DE by the diameter AB and each line parallel to AB are right angles, all the parts of the lines of the semicircle DAE will fall upon and coincide with those of the other semicircle; that is, CA with CB; FL with LG; and HK with KI; also MD with DN and PE with EQ; therefore ;

1st. The chords are bisected.

2d. The two lines MN and PQ can only touch the circumference at D and E respectively; for, at these points, the straight lines and the curves may be regarded as starting and taking different directions, for the straight lines are parallel to the chords, whilst the curves approach and intersect them.

3d. As the one half of each chord falls upon its other half, and the one side of each tangent falls upon its other side, so also the intercepted arcs respectively fall upon and coincide with each other, and hence are equal.

COR. 1.—A radius perpendicular to a chord bisects the chord and also its arc.

Cor. 2.—A tangent is perpendicular to a radius at its extremity.

Cor. 3.—A line perpendicular to a chord at its middle point, passes through the center of the circle.

Scho.—Observe that a tangent touches the circumference at but one point, and a chord intersects the circumference at two points, each end passing away in opposite directions ; hence a straight line can only intersect a circumference at two points.

THEOREM IX.

If a perpendicular be erected at the middle point of a straight line, every point in the perpendicular is equally distant from the extremities of the line.

Let PC be perpendicular to AB at its middle point C; then will any point in the line PC be equally distant from A and B.

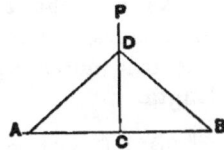

Take any point in the perpendicular . PC as D, and draw AD and BD ; and let the part ACD be revolved on DC as an axis, until it fall upon the plane of BCD; since both angles at C are right angles, CA will take the direction of CB, and as AC is equal to CB, the point A will fall upon the point B, and CA will coincide with CB, and AD must fall upon and coincide with DB. (Special Ax. 2.)

Cor. 1.—As two points determine the direction of a line; any straight line which has two points equally distant from the extremities of another line, is perpendicular to the latter at its middle point.

Cor. 2.—With D as a center and DA as a radius, a circum ference may be described which will pass through the points A and B, and AB becomes a chord of the circumference; and as a straight line cannot intersect a circumference at more than two points, there can be only two points in the line AB equally distant from the point D.

THEOREM X.

A perpendicular is the shortest distance from the center of a circle to a chord, or from a point to a line.

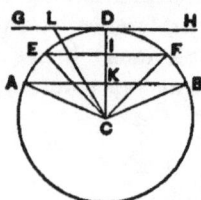

Let AB and EF be two parallel chords, and draw CD perpendicular to AB; it will also be perpendicular to EF. At the point D, the extremity of the radius CD, draw GH perpendicular to CD, and it will be parallel to the chords AB and EF.

1st. The perpendicular CK is less than CA or CB, each of which is equal to CD, of which CK is a part.

CI is less than CE or CF for the same reason.

As the chord departs from the center and consequently diminishes, the perpendicular approaches the radius in length, but can never equal it whilst the chord has any definite length.

2d. CD is less than any oblique line drawn from the point C to GH; for any oblique line as CL will terminate without the circumference, and consequently be greater than the radius.

COR. 1.—A perpendicular is the shortest distance from a point to a line, and also between two parallels.

COR. 2.—The farther distant from the center, the less the chord.

COR. 3.—The less the chord the less the arc, and consequently the less the opposite angle.

PROBLEM I.

To bisect a given line.

Let AB be the given line; then with A and
B as centers, and a radius greater than the
half of AB, describe arcs above and below the
line AB, intersecting at D' and E, and join
D' and E, cutting AB in C, which will be the
middle point. (Th. 9, Cor. 1.)

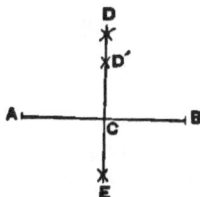

SCH. 1.—The intersections may both be
made on the same side of AB, as at D' and D,
by taking different radii.

SCH. 2.—As the radius becomes, as it were, an oblique line,
whilst one-half of AB is a perpendicular, it must, of course, be
greater than one-half of AB.

PROBLEM II.

*From a point without a line, to draw a perpendicular
to the line.*

Let P be a point without the line CD.
With P as a center, and a radius greater
than the shortest distance to CD, which
would be a perpendicular, draw an arc cutting CD in A and B; then with A and B as
centers and a radius greater than one-half of
AB, describe arcs intersecting at E; then
will P and E be two points equally distant from A and B, and
hence PE is perpendicular to AB or to CD. (Th. 9, Cor. 1.)

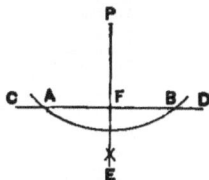

PROBLEM III.

At a point in a line, to erect a perpendicular to the line.

Let P be a point in the line CD; then,
with P as a center and a radius PA, cut CD
in two points, A and B; and with A and B
respectively as centers, and a radius greater
than one-half of AB, describe arcs intersect-
ing at E, and join PE. It will be perpendicular to CD at the
point P.

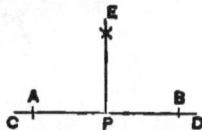

PROBLEM IV.

To bisect a given angle.

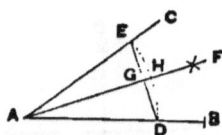

Let BAC be the given angle. Then with A as a center and a radius that will cut the sides AB and AC, draw the arc DE and its chord; then with D and E, respectively, as centers and a radius greater than the half of DE draw arcs intersecting at F, and join AF. The two points A and F will be equally distant from D and E; hence the line AF will bisect the chord DE, its arc, and hence the angle A.

PROBLEM V.

From a point without a line, to draw a parallel to the line.

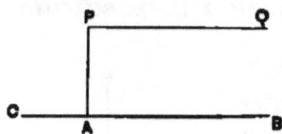

Let P be a point without the line CB. From P draw PA perpendicular to CB (Prob. 2), and from P draw a perpendicular to AP. Then will PQ be parallel to CB. (Th. 5.)

PROBLEM VI.

To find the center of a given circle.

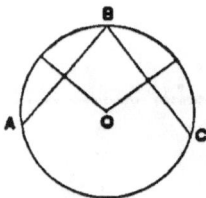

Draw any two chords, as AB and BC, to the given circle, and pass perpendiculars through their middle points; both perpendiculars will pass through the center of the circle. (Th. 8, Cor. 3.) The point of their intersection O, which is the only common point, is the center of the circle.

Scho.—It is not necessary that the chords be consecutive, but they must not be parallel, as then there would be but one perpendicular and the same perpendicular would pass through the middle points of both chords.

PROBLEM VII.

To circumscribe a circle about a triangle.

Let ABC be the given triangle. Pass
the perpendiculars DO and EO through
the middle points of any two sides of the
triangle, as AC and CB; their point of
intersection O will be the center of a circle
of which AC and CB are chords.

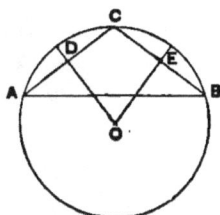

Cor. 1.—A circumference can be passed
through any three points not in the same straight line; but if
the three points are in the same straight line, only one perpendicular can be drawn, and hence no solution.

Cor. 2.—Each of the three points is equally distant from the
center, but only two points in the same straight line can be
equally distant from a point without the line.   (Th. 9, Cor. 2.)

BOOK II.

DEF.—An **Inscribed Angle** has its vertex in the circumference of a circle of which its sides are chords.

THEOREM I.

An inscribed angle is measured by one-half the arc intercepted by its sides.

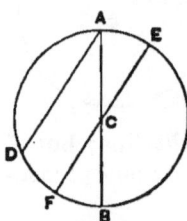 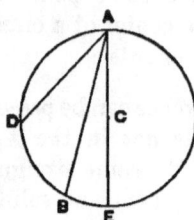 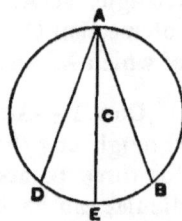

FIRST CASE.   SECOND CASE.   THIRD CASE.

There are three cases:

1st. When one of its sides is a diameter; as, the angle BAD has one side AB a diameter. Through the center C draw EF parallel to the chord AD; then will the angles BAD and BCF be equal, as they are corresponding angles; but BCF and ACE are equal, as they are vertical angles—they are both angles at the center, measured respectively by the arcs FB and AE, which arcs are consequently equal. Therefore arc AE is equal to arc FB, also equal to arc DF; hence FB is one-half of BD. Therefore the angle A is measured by ¼ arc BD, which is the arc intercepted by its sides.

2d. When the center of the circle is without the angle, as the angle BAD.

Here   angle EAD is measured by ½ arc DE,
And   "   BAE "   "   " ½ " BE.
By subtraction, "   BAD "   "   " ½ " BD.

3d. When the center is within the triangle; as BAD.
Here   angle BAE is measured by ½ arc BE,
And   "   DAE "   "   " ½ " DE.
By addition,   "   BAD "   "   " ½ " BD.

THEOREM I.—CONTINUED.

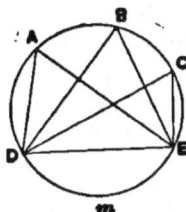

Cor. 1.—All the angles inscribed in the same segment are equal; since the angles A, B, and C are measured each by the half of the same arc D*m*E, they are equal.

Cor. 2.—An angle inscribed in a semicircle is a right angle, as BAD.

Cor. 3.—An angle inscribed in a segment greater than a semicircle is acute, as BAC; and an angle inscribed in a segment less than a semicircle is obtuse, as BDC.

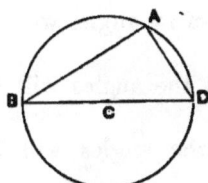
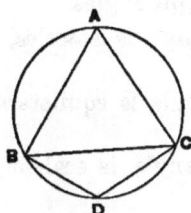
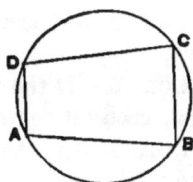

Cor. 2.   Cor. 3.   Cor. 4.

Cor. 4.—The opposite angles of an inscribed quadrilateral are supplementary; as,

Angle A measured by ½ arc BCD.
"   C   "   " " " DAB.

The sum of the two arcs is the circumference; hence half their sum is the measurement of two right angles.

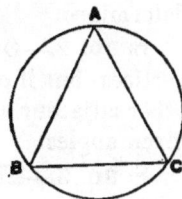

Cor. 5.—If the extremities of the chords forming an inscribed angle be joined by a straight line, an inscribed triangle is formed. As in the angle A join B and C; then the triangle ABC has each of its angles inscribed, and is therefore an inscribed triangle.

THEOREM II.

The sum of the three angles of any triangle is equal to two right angles.

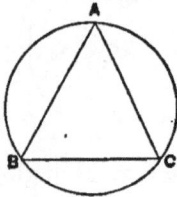

Let ABC be an inscribed triangle; then will angles A + B + C equal two right angles; as,

The angle A is measured by ½ arc BC,
"   " B "   "   " " " AC,
"   " C "   "   " " " AB.

The sum of the three arcs is a circumference, one-half of which measures the angles and is the measure of two right angles; hence, the sum of the angles of an inscribed triangle is equal to two right angles; but, as a circumference may be passed through all the vertices of any triangle and the triangle become inscribed, (Book I, Problem 7), it follows that the sum of the three angles of any triangle is two right angles.

COR. 1.—If the triangle is isosceles, two of its angles will be equal.

COR. 2.—If the triangle is equilateral, all the angles will be equal, each 60 degrees.

COR. 3.—If the triangle is scalene, all the angles will be unequal.

COR. 4.—As an inscribed angle is measured by half the arc intercepted by its sides, and the greater the arc the greater the chord, hence the greatest angle is opposite the greatest chord and the next to the greatest angle opposite the chord next to the longest, and the smallest angle opposite the shortest chord; consequently, in any triangle the greatest angle is opposite the longest side, the next to the greatest angle opposite the next to the longest side; and the smallest angle opposite the shortest side.

SCHO. 1.—If a triangle has two sides and the included angle given, the three vertices of the triangle are fixed, and the triangle determined.

SCHO. 2.—One side and the two adjacent angles fix the three vertices, but if one of the given angles be the opposite angle, the other adjacent angle is the supplement of the sum of the two given angles.

SCHO. 3.—The three sides of a triangle also determine the triangle.

COR.—Two triangles, each having the three parts named in either of the scholia respectively equal, are equal in all their parts.

REM.—The sum of the angles of a triangle may be determined by means of the parallels, as follows:

Let ABC be any triangle. At C draw DE. parallel to AB. The angles marked 1 and 1 and 2 and 2 are respectively alternate interior angles, and consequently respectively equal; hence, the three angles of the triangle are equal to all the angles at a point on one side of a straight line, which is two right angles. (Book I, Th. 3, Cor. 2.)

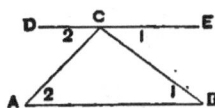

THEOREM III.

If one side of a triangle is produced in one direction, the exterior angle formed is equal to the sum of the two interior angles not adjacent.

Let ABC be a triangle. Produce BC to D, forming the exterior angle ACD. From C draw CE parallel to BA; then will the angles marked 1 and 1 be corresponding angles and equal, and the angles marked 2 and 2 be alternate interior angles and equal; hence, the exterior angle ACD is equal to the sum of ABC and BAC, the two interior angles not adjacent.

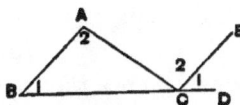

THEOREM IV.

Every point in the line which bisects an angle is equally distant from each side of the angle.

Let AP bisect the angle A, and revolve the part CAP on AP as an axis; AC will fall upon and coincide with AB, since angle CAP is equal to angle PAD. From any point as P draw PD perpendicular to AB; it will be the shortest distance to AB, and also to AC, which coincides with AB.

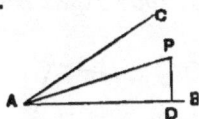

THEOREM V.

If from a point without a line a perpendicular be drawn to the line, and oblique lines to different points of the line:

1st. The perpendicular will be shorter than any oblique line.

2d. Any two oblique lines at equal distances from the foot of the perpendicular will be equal.

3d. The farther from the foot of the perpendicular, the greater the oblique line.

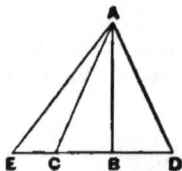

1st. In the triangle ABD, the angle B is a right angle; hence it is greater than the angle D. Therefore the side AB opposite the angle D, is less than the side AD opposite the angle B.

2d. The triangles ABD and ABC have each two sides and the included angle respectively equal; hence the triangles are equal, and AC equal to AD.

3d. In the triangle ACE, the angle ACE is obtuse, consequently greater than the angle AEC; therefore the side AE is greater than the side AC.

THEOREM VI.

The sum of all the angles of any quadrilateral is equal to four right angles.

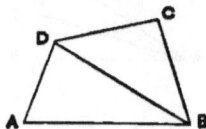

Let ABCD be any quadrilateral. Draw the diagonal DB, dividing the quadrilateral into two triangles. All the angles of the two triangles make up precisely the angles of the quadrilateral; but, the sum of all the angles of the two triangles is four right angles. Hence, the sum of all the angles of any quadrilateral is four right angles.

THEOREM VII.

The sum of all the angles of any polygon is equal to two right angles taken as many times as the polygon has sides, minus four right angles.

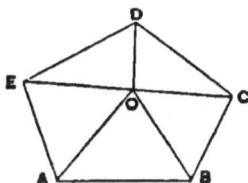

Let ABCDE be any polygon. Take any point within the polygon, as O, and from it draw lines to the extremities of all the sides. The number of triangles will be equal to the number of sides of the polygon. The sum of the angles of each triangle is two right angles; hence, the sum of all the angles of the triangles which make up the polygon, is two right angles taken as many times as the polygon has sides; but all the angles at O, which equal four right angles, belong to the triangles, but not to the polygon, and must be deducted from the sum of all the angles of the triangles, and the difference will be the angles of the polygon; therefore, the sum of all the angles of any polygon is equal to two right angles taken as many times as the polygon has sides minus four right angles.

THEOREM VIII.

If each side of a polygon is prolonged, the sum of all the exterior angles thus formed will be equal to four right angles.

At each vertex of the polygon, the sum of the interior and exterior angles is two right angles; hence, the sum of all the interior and exterior angles is equal to two right angles taken as many times as the polygon has sides, which sum is four right angles more than the sum of all the interior angles. (Th. 8.) Therefore, the sum of all the exterior angles is four right angles.

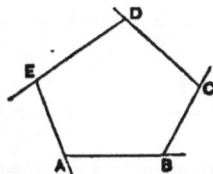

2

THEOREM IX.

The opposite sides and opposite angles of a parallelogram are respectively equal.

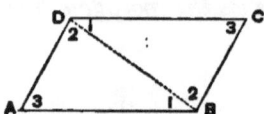

Let ABCD be a parallelogram. Draw the diagonal DB. Since AB and DC are parallels cut by DB, the angles 1 and 1 are alternate and equal; and since AD and BC are parallels cut by DB, the angles 2 and 2 are alternate and equal. The triangles ABD and BCD have the side BD common and the adjacent angles equal; hence the triangles are equal. Therefore, AB is equal to DC, and AD equal to BC. The angles 3 and 3 are equal, and the sums of the same angles at B and D are equal.

THEOREM X.

If the opposite sides of a quadrilateral are respectively equal, it will be a parallelogram.

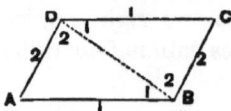

Let AB equal DC, and AD equal BC. Draw the diagonal DB. The triangles ABD and BCD have their three sides respectively equal; hence the triangles are equal, and the angles opposite the equal sides equal, that is, 1 equals 1, and 2 equals 2, and these are respectively alternate angles; hence, the opposite sides are parallel, and the quadrilateral is a parallelogram.

THEOREM XI.

If two opposite sides of a quadrilateral are equal and parallel, the figure will be a parallelogram.

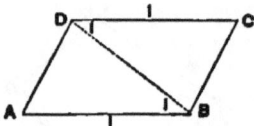

Let AB and DC be equal and parallel, and draw the diagonal DB.

Since AB and DC are parallel, the alternate angles 1 and 1 are equal, and the triangles ABD and DCB have respectively two sides and an included angle equal, and are therefore equal, which makes the other sides equal and parallel, and the opposite angles of the figure equal; hence it is a parallelogram.

THEOREM XII.

The diagonals of a parallelogram mutually bisect each other.

The triangles ABE and DCE have a side and two adjacent angles respectively equal; hence the side AE opposite the angle 2, is equal to EC opposite the angle 2 in the other triangle, and DE is equal to EB for the same reason.

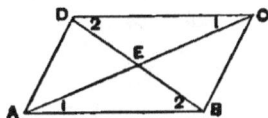

THEOREM XIII.

An angle formed by a tangent and a chord is measured by one-half the intercepted arc.

The angle BAD, formed by the tangent AD and the chord AB, is measured by ½ arc AmB.

From B draw the chord BE parallel to the tangent AD, then the angles BAD and ABE are alternate interior angles and consequently equal. The angle ABE is inscribed, and is measured by ½ arc AnE, which is equal to the arc AmB, as they are intercepted by two parallels; consequently the angle BAE is measured by ½ arc AmB.

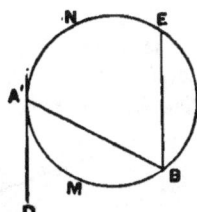

THEOREM XIV.

An angle formed by two chords intersecting within the circle, is measured by one-half the sum of the intercepted arcs.

Let AB and DE be two chords intersecting at C.

From A draw the chord AF parallel to DE, then will the angles BAF and BCE be corresponding angles, and equal; the angle BAF is inscribed, and is measured by ½ arc BEF; but the arc FE is equal to the arc AD, therefore arc BEF is equal to the sum of the arcs EB and AD; consequently the angle BCE, or its equal ACD, is measured by ½ the sum of the arcs included by its sides.

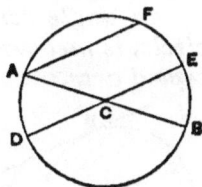

THEOREM XV.

An angle formed by two secants meeting without the circle, is measured by one-half the difference of the intercepted arcs.

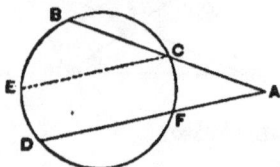

The angle A is formed by the two secants AB and AD.

From C draw CE parallel to AD; BAD and BCE are corresponding angles. The angle BCE is measured by $\frac{1}{2}$ arc BE = BD — CF; therefore, the angle A, formed by two secants meeting without the circle, is measured by one-half the difference of the intercepted arcs.

THEOREM ·XVI.

An angle formed by a tangent and a secant meeting without a circle, is measured by one-half the difference of the intercepted arcs.

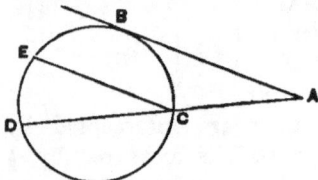

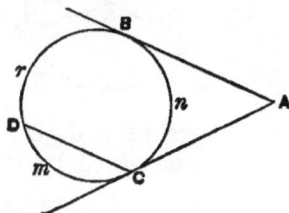

The angle BAD is formed by a tangent and a secant meeting at A. From C draw CE parallel to AB; then the angles BAD and DCE are corresponding and equal; but the angle DCE is inscribed, and measured by $\frac{1}{2}$ arc DE, and DE = DB — BE, or its equal BC; parallels intercept equal arcs, therefore the angle A is measured by $\frac{1}{2}$ arc DE = $\frac{1}{2}$ arc (DB — BC).

THEOREM XVII.

An angle formed by two tangents meeting without a circle, is measured by one-half the difference of the intercepted arcs.

The angle A is formed by two tangents meeting without the circle. At C draw CD parallel to AB, then the angles BAC and DCE will be corresponding and equal; but the angle DCE is formed by a tangent and a chord, and hence is measured by $\frac{1}{2}$ arc C*m*D. (Th. 14.) Arc D*m*C = (arc B*r*D*m*C — arc B*r*D) and arc B*r*D = arc B*n*C ; therefore, an angle formed by two tangents intersecting without a circle is measured by one-half the difference of the intercepted arcs.

THEOREM XVIII.

The side of a regular hexagon is equal to the radius of the circumscribed circle.

Describe a circumference and make the chord AB equal to the radius of the circle. Draw the radii CA and CB. CAB will be an equilateral triangle, each side equal to the radius of the circle, and each angle equal to 60 degrees ; hence, AB is a chord of an arc of sixty degrees, which is contained exactly six times in the circumference, and is therefore a side of a regular hexagon.

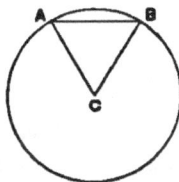

PROBLEM I

To construct an angle equal to a given angle.

Let A be the given angle.

With A as a center and a radius that will cut both sides, describe the arc CD ; then, with the same radius and B as a center, describe the arc EF, making it equal to CD, and draw BF ; and EBF will be the required angle.

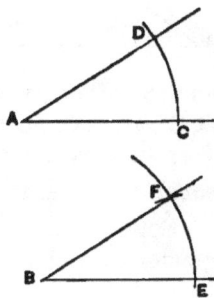

PROBLEM II.

Two sides and the included angle given, to construct a triangle.

Make the angle A equal to the given angle, and on one side lay off AB equal to one of the given sides, and on the other AC equal to the other given side. Draw BC, and ABC will be the required triangle.

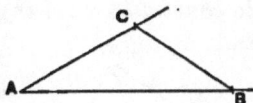

Cor.—Two triangles having two sides and the included angle respectively equal, are equal in all their parts.

PROBLEM III.

One side and the two adjacent angles given, to construct the triangle.

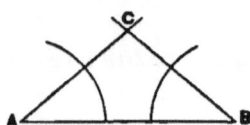

Make AB equal to the given side. At A construct an angle equal to one of the given angles, and at B an angle equal to the other given angle; the intersection C of the lines forming these angles will be the vertex of the third angle, and ABC will be the required triangle.

COR.—Two triangles having each a side and the two adjacent angles respectively equal, are equal in all their parts.

PROBLEM IV.

To construct a triangle, having given the three sides.

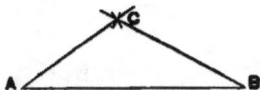

Make AB equal to one of the given sides. Then with A as a center, and a radius equal to one of the given sides, describe an arc; and with B as a center and the other given side, describe an arc intersecting the other arc at C and draw AC and BC, then ABC will be the required triangle.

COR.—Two triangles having their three sides respectively equal, are equal in all their parts.

SCHO.—The sum of any two sides of a triangle must be greater than the third side.

PROBLEM V.

Two sides and an angle opposite one of them given, to construct a triangle.

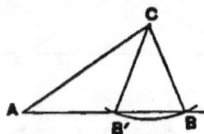

Make the angle A equal to the given angle, and make AC equal to one of the given sides; then with C as a center and a radius equal to the other given side, draw the arc BB', and draw CB and CB'. In this case there are two triangles.

SCHO.—The second side must be equal to or greater than the perpendicular from C to AB. If it is equal, there will be one right-angled triangle; if it be greater than the perpendicular and less than CA, there will be two triangles; but if it be less than the perpendicular, there will be no triangle.

PROBLEM VI.

Form an equilateral triangle.

Describe a circle, and apply the radius six times to the circumference, and draw the chords; the result is a hexagon. Join the alternate vertices, and the result is ABC, an equilateral triangle.

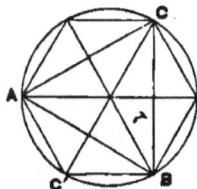

PROBLEM VII.

To construct a regular polygon of eight sides.

Describe a circumference, and divide it into eight equal parts. Draw chords to the equal arcs; they will be the sides of the polygon.

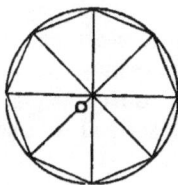

Draw radii from the extremities of the sides to the center of the circle; there will be as many isosceles triangles as the polygon has sides. The angles at the center are equal, having equal arcs; and each angle of the polygon is composed of two equal angles of the isosceles triangles; hence, all the angles of the polygon are equal; and the sides being also equal, the polygon is regular.

COR.—A regular polygon of any number of sides may be constructed by dividing the circumference into as many equal parts as there are sides.

REM.—The circumference will be divided into eight equal parts by applying the chord of an arc of 45°.

PROBLEM VIII.

To draw a tangent to the circumference at any point on it.

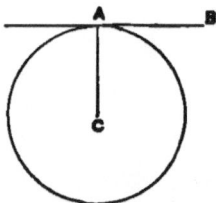

Let C be the center of a given circle, and A any point in its circumference. Draw the radius CA, and from A draw AB perpendicular to the radius CA; then AB will be the tangent required. (Book 1, Th. 8, Cor. 2.)

PROBLEM IX.

From a point without the circle, to draw a tangent to the circle.

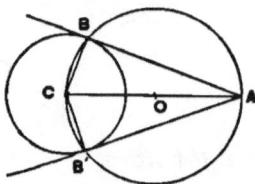

Let C be the center of the given circle, and A the point without the circle from which the tangent is to be drawn. Join the point A and the center C and bisect AC in O; then, with O as a center and the radius OC describe a circumference; the points B and B', the intersections of the two circumferences, will be the points of tangency, AB and AB' the tangents, as each is a perpendicular to a radius at its extremity; the angles ABC and AB'C being each inscribed in a semicircle.

PROBLEM X.

On a straight line, to construct a segment that shall contain a given angle.

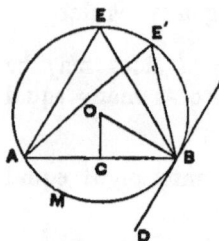

Let AB be the given line. At B make the angle ABD equal to the given angle. Draw BO perpendicular to BD, and at C, the middle point of AB, erect a perpendicular intersecting BO at O; then, with O as a center and radius OB, describe a circumference to which DB is a tangent and AB a chord, and the angle ABD is measured by ½ arc AmB; so also every angle, as E, E', inscribed in the segment AEB.

PROBLEM XI.

To inscribe a circle in a given triangle.

Bisect any two angles as A and B by the straight lines AD and BD, and as every point in each bisecting line is equally distant from the sides of the angle, hence the point of intersection D will be equally distant from the three sides, and DE, DF and DG will be radii of the inscribed circle.

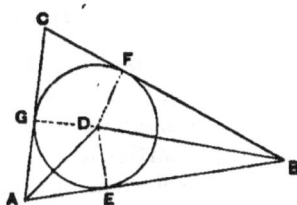

PROBLEM XII.

Draw a common tangent to two external circles of different radii.

Let C and C' be the centers of two circles which are external. With C as a center and a radius equal to the difference of the radii of the circles, describe a small circumference, and from the point C' draw a tangent C'A' to this small circumference; from the center C draw a radius through the point of tangency A', and extend it to A in the circumference of the large circle; draw C'B parallel to CA and join AB, which will be the required tangent.

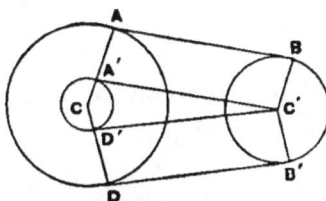

COR.—A second tangent B'D may always be drawn.

PROBLEM XIII.

To draw a tangent to two external circumferences of different radii, the tangent passing between the circles and touching at points on the opposite sides of the circumferences.

Let CA and C'B be the radii of the given circles. With C as a center and a radius equal to the sum of the two given radii, draw a circumference. From C' draw C'D and C'D' tangents to the large circle; then draw the radii CD and CD', cutting the circumference of the smaller given circle in A and A', which will be the points of tangency. From C' draw C'B parallel to CA, and C'B' parallel to CA', and join AB and A'B', and they will be the required tangents.

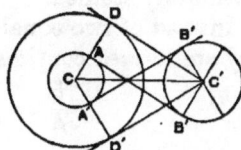

BOOK III.

PROPORTIONS.

DEFINITION.

When two quantities, each having the form of a fraction, that is, each having a numerator and a denominator, are equal to each other, an equation may be formed of them; and they may be arranged proportionally. In order to show whether the ratio is increasing or decreasing, the denominators should be made the antecedents and the numerators consequents; still, they are in proportion when taken in an inverse order, that is, the numerators as antecedents and the denominators as consequents, but the ratios will be inverted. Thus,

$$\frac{B}{A} = \frac{D}{C}.$$

Then will $\qquad$ A : B :: C : D,

and $\qquad$ B : A :: D : C.

This proportion is made very simple by an arithmetical solution; thus,

$$\frac{10}{5} = \frac{12}{6}.$$

By reduction, $\qquad \dfrac{2}{1} = \dfrac{2}{1}$,

and $\qquad$ 1 : 2 :: 1 : 2.

The same proportion as 5 : 10 :: 6 : 12,

or $\qquad$ 2 : 1 :: 2 : 1,

and $\qquad$ 10 : 5 :: 12 : 6.

The equation is true if the fractions are inverted; thus,

$$\frac{5}{10} = \frac{6}{12}, \quad \text{and} \quad \frac{1}{2} = \frac{1}{2}; \quad \therefore \quad 2 : 1 :: 2 : 1.$$

Proportions are much used in Geometry, and should therefore be carefully studied.

Instead of two equal ratios there may be many, in which case they are termed continued proportions; as,

$$\frac{B}{A} = \frac{D}{C} = \frac{F}{E} = \frac{H}{G} = \frac{K}{I}, \text{ etc.}$$

Which may be rendered,

$$A : B :: C : D :: E : F :: G : H :: I : K, \text{ etc.}$$

This is read: as A is to B, so is C to D, so is E to F, so is G to H, so is I to K. The antecedent and consequent form a couplet, and in a continued proportion any two couplets, may be taken to form a proportion of four terms, which is always considered a proportion, and the first and last terms are called extremes, and the second and third the means.

THEOREM I.

If four quantities are proportional, the product of the means equals that of the extremes.

If $$A : B :: C : D,$$

then $$\frac{B}{A} = \frac{D}{C}.$$

Clearing of fractions or multiplying both members by A and C (Gen. Ax. 6), $$BC = AD.$$

COR. 1.—B : A :: D : C; that is, if four quantities are in proportion, they are also in proportion by inversion.

COR. 2.—They are also in proportion by alternation ; thus, $\frac{B}{A} = \frac{D}{C}.$ Multiplying both members by $\frac{C}{B}$, $\frac{BC}{AB} = \frac{CD}{BC}$; reducing, $\frac{C}{A} = \frac{D}{B}$; therefore, A : C :: B : D, and again by inversion,

$$C : A :: D : B.$$

THEOREM II.

A mean proportional between two quantities is equal to the square root of their product.

Let B be a mean proportional between A and C ; as,

$$A : B :: B : C.$$

The product of the means is equal to that of the extremes; thus,

$$B^2 = A \times C.$$

Extracting the root of both members,

$$B = \sqrt{A \times C}.$$

THEOREM III.

If the product of two quantities equals the product of two other quantities, either of the two forming a product may be made the means, and the other two the extremes of a proportion.

Let $$B \times C = A \times D;$$
divide by $A \times C$, then

$$\frac{B \times C}{A \times C} = \frac{A \times D}{A \times C} = \frac{B}{A} = \frac{D}{C},$$

and $$A : B :: C : D, \tag{1}$$

or $$C : D :: A : B. \tag{2}$$

In the first proportion, A and D are the extremes, and B and C the means; in the second, B and C are the extremes, and A and D the means.

THEOREM IV.

If four quantities are proportional, they will also be proportional by composition and division.

If $\dfrac{B}{A} = \dfrac{D}{C}$, then $\dfrac{B}{A} + 1 = \dfrac{D}{C} + 1$, and $\dfrac{B}{A} - 1 = \dfrac{D}{C} - 1$.

Reducing to improper fractions,

$$\frac{B + A}{A} = \frac{D + C}{C}, \quad \text{and} \quad \frac{B - A}{A} = \frac{D - C}{C},$$

and $\quad A : B+A :: C : D+C;\quad$ also, $\quad A : B-A :: C : D-C.$

By alternation,

$$A : C :: B+A : D+C;\quad \text{also,} \quad A : C :: B-A : D-C.$$

$$\frac{C}{A} = \frac{D + C}{B + A};\quad \text{also,} \quad \frac{C}{A} = \frac{D - C}{B - A}.$$

Gen. Ax. 1, $\quad \dfrac{D+C}{B+A} = \dfrac{D-C}{B-A};\quad \therefore\ B+A : D+C :: B-A : D-C.$

By alternation, $\quad B + A : B - A :: D + C : D - C.$

THEOREM V.

Like powers and like roots of proportional quantities are proportional.

Squaring both sides, $\dfrac{B}{A} = \dfrac{D}{C}$; then $\dfrac{B^2}{A^2} = \dfrac{D^2}{C^2}$,

and $\qquad \dfrac{B^n}{A^n} = \dfrac{D^n}{C^n}$, and $\qquad \dfrac{B^{\frac{1}{n}}}{A^{\frac{1}{n}}} = \dfrac{D^{\frac{1}{n}}}{C^{\frac{1}{n}}}$. (Gen. Ax. 10.)

$\therefore \quad A^2 : B^2 :: C^2 : D^2$, and $\quad A^n : B^n :: C^n : D^n$,

and $\qquad\qquad A^{\frac{1}{n}} : B^{\frac{1}{n}} :: C^{\frac{1}{n}} : D^{\frac{1}{n}}$.

THEOREM VI

Any equimultiple of one couplet will be proportional to the other couplet or to any equimultiple of it.

This depends upon the principle that multiplying both numerator and denominator of a fraction by the same quantity does not change its value.

$$\frac{B}{A} = \frac{D}{C}, \quad \text{and} \quad \frac{mB}{mA} = \frac{D}{C} = \frac{nD}{nC}.$$

THEOREM VII.

If the corresponding terms of two proportions be multiplied, their products will be proportional.

$$A : B :: C : D$$
$$E : F :: G : H$$
$$\frac{B}{A} = \frac{D}{C},$$
$$\frac{F}{E} = \frac{H}{G};$$

hence (Gen. Ax. 6), $\qquad \dfrac{BF}{AE} = \dfrac{DH}{CG}.$

$\therefore \quad AE : BF :: CG : DH.$

THEOREM VIII.

In a series of proportions, as one antecedent is to its consequent, so is the sum of all the antecedents to the sum of all the consequents.

$$A : B :: C : D :: E : F :: G : H, \text{etc.}$$

$$AD = BC$$
$$AF = BE$$
$$AH = BG$$
$$\underline{AB = BA}$$
$$A(B + D + F + H) = B(A + C + E + G)$$

$$\therefore \quad A : B :: A + C + E + G : B + D + F + H.$$

Cor. 1.—If any two proportions have an equal ratio, then the other terms are proportional.

Cor. 2.—The same is true if the antecedents are the same in two proportions.

BOOK IV.

THEOREM I.

The area of a rectangle is equal to the product of its base and altitude.

There may be three cases:

1st. When the base and altitude are composed of units of the same denomination; then it is evident that there will be as many square units for every unit in altitude as there are units in the base; and for every additional unit in altitude as many more square units; hence the area will be the product of the base and altitude.

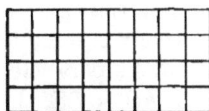

2d. If there be a fraction in one or both the dimensions, the common denominator will be the denomination of the unit of measure; hence, the product of the base and altitude will give the area, in units of the same denomination.

3d. If the dimensions are incommensurable, the unit of measure will be an infinitesimal.

THEOREM II.

The area of a parallelogram is equal to the product of its base and altitude.

Let ABCD be a parallelogram, AB its base, BE its altitude, its area $= AB \times BE$. Construct the rectangle ABEF; its area $= AB \times BE$. FE $=$ AB and DC $=$ AB, $\therefore$ FE $=$ DC, and taking from each DE, there remains FD $=$ EC; hence the triangles ADF and BCE are equal, having all their sides equal. In changing the parallelogram into the rectangle, we have added and subtracted the same area; hence the parallelogram is equal to the rectangle. $\therefore$ the area of the parallelogram is $AB \times BE$, product of base and altitude.

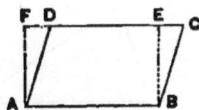

THEOREM III.

The area of a triangle is equal to one-half the product of the base and altitude.

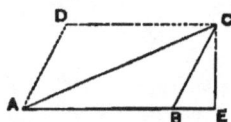

Area ABC $= \frac{1}{2}$ (A B $\times$ CE).

Let ABC be the given triangle, AB its base, and EC its altitude. Construct a parallelogram on AB as one of its sides and BC as another, draw AD parallel to BC and CD parallel to AB; then will ABCD be a parallelogram. The triangles ABC and ACD will have their three sides respectively equal ; hence the triangles are equal and each is one-half of the parallelogram ABCD; and as the area of the parallelogram is AB $\times$ CE, that of the triangle is $\frac{1}{2}$ (AB $\times$ CE); therefore, the area of a triangle is equal to one-half the product of the base and altitude.

COR. 1.—Rectangles, parallelograms and triangles are to each other as the products of their bases and altitudes respectively.

COR. 2.—If the bases are equal, they are to each other as their altitudes.

COR. 3.—If the altitudes are equal, they are to each other as their bases.

THEOREM IV.

The area of a trapezoid is equal to the product of its altitude and half the sum of its parallel bases.

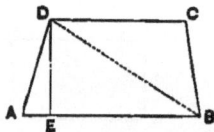

Let ABCD be a trapezoid, DE its altitude, and AB and DC its parallel bases. Draw the diagonal DB, dividing the trapezoid into two triangles whose common altitude is DE and their bases AB and DC.

The area of the triangle ABD $= \frac{1}{2}$ (AB $\times$ DE),

" " " " " BCD $= \frac{1}{2}$ (DC $\times$ DE),

By addition, area of ABCD $=$ DE $\frac{1}{2}$ (AB $+$ DC).

That is, the area of a trapezoid is equal to the product of its altitude and $\frac{1}{2}$ the sum of its bases.

THEOREM V.

The square described on the sum of two lines, is equivalent to the sum of the squares of the lines, increased by twice the rectangle of the lines.

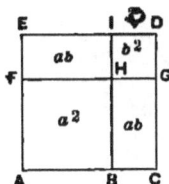

ACDE is the square described on the sum of AB and BC; and corresponds to the algebraic formula $(a + b)^2 = a^2 + 2ab + b^2$, in which AB $= a$ and BC $= b$.

Cor.—If the lines are equal there will be four equal squares. Let AB $= 1$ and BC $= 1$; then the square of two is four times the square of one.

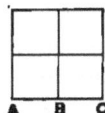

THEOREM VI.

The square described on the difference of two lines, is equivalent to the sum of the squares of the lines, diminished by twice the rectangle of the lines.

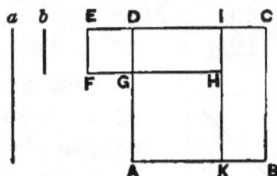

$$AB = a, \qquad KB = b,$$
$$(a - b)^2 = a^2 - 2ab + b^2,$$
$$ABCD = a^2, \qquad EDFG = b^2,$$
$$BCIK = ab, \quad \text{and} \quad EIFH = ab,$$

THEOREM VII.

The rectangle contained by the sum and difference of two lines, is equivalent to the difference of their squares.

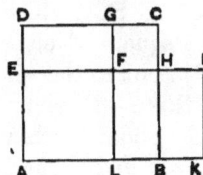

$$AB = a, \qquad \text{and} \qquad LB = BK = b,$$
$$a + b = AK, \qquad \text{and} \quad a - b = AE = AL,$$
$$(a + b) \times (a - b) = a^2 - b^2,$$
$$ABCD = a^2, \qquad \text{and} \qquad FHGC = b^2,$$

The rectangle EFGD = rect. BKIH = $b (a - b)$,
" " ABHE = $= a (a - b)$,
By addition, AKIE = $= (a + b) (a - b)$.

THEOREM VIII.

The square described on the hypothenuse of a right-angled triangle, is equivalent to the sum of the squares of the other two sides.

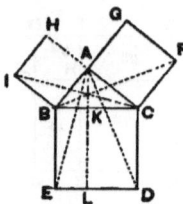

Let ABC be a triangle, right-angled at A, then will $\overline{BC}^2 = \overline{AB}^2 + \overline{AC}^2$,

Construct a square on each side of the triangle. From A draw a perpendicular to BC and extend it to ED, and draw AE, AD, IC and BF. The triangles ABE and IBC have two sides respectively equal, viz., AB = BI and BC = BE, being respectively sides of the same square, and the included angles equal; that is, ABE = IBC, as each one is composed of the angle ABC and a right angle; hence, triangle ABE = triangle IBC; but triangle ABE is one-half the rectangle BELK, having the same base and altitude BE and BK; and the triangle IBC is one-half the square ABIH = IB × AB = $\overline{AB}^2$; therefore, $\overline{AB}^2$ = rectangle BELK.

By the same process, we prove the triangle BCF = ACD, and the square ACFG = rect. CDLK ; therefore, $\overline{BC}^2 = \overline{AB}^2 + \overline{AC}^2$. And by transposing,

COR. 1. $\overline{BC}^2 - \overline{AB}^2 = \overline{AC}^2$, and $\overline{BC}^2 - \overline{AC}^2 = \overline{AB}^2$.

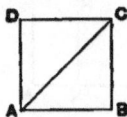

COR. 2.—The square described on the diagonal of a square is double the square described on the side, as the sides are equal; hence, the square of diag. : sq. of side :: 2 : 1, and diag. : side :: $\sqrt{2}$: 1.

COR. 3. Since $\overline{AB}^2$ = rect. BELK, and $\overline{AC}^2$ = rect. CDLK, the resulting proportion $\overline{AB}^2 : \overline{AC}^2 :: BK : KC$; that is, the squares of the sides are proportional to their adjacent segments of the hypothenuse. And $\overline{BC}^2 : \overline{AB}^2 :: BC : BK$, and $\overline{BC}^2 : \overline{AC}^2 :: BC : KC$; that is, the square of the hypothenuse is to the square of either side as the hypothenuse is to the segment adjacent to the side.

SCHO.—Observe, that if the right angle A be diminished, the sides about it remaining the same, the third side BC will be diminished; and if the angle A be increased, BC will be increased; in the first case the square of BC will be less, and in the second greater than the sum of the other two ; hence, the right-angled triangle is the only one in which the square of one side is equivalent to the sum of the squares of the two.

THEOREM IX.

In any triangle, the square of a side opposite an acute angle is equivalent to the sum of the squares of the two other sides, minus twice the rectangle of the base, and the distance from the acute angle to the foot of the perpendicular let fall from the vertical angle on the base, or the base produced.

In the triangle ABC the side AB is opposite the acute angle C; hence,

$$\overline{AB}^2 = \overline{AC}^2 + \overline{BC}^2 - 2BC \times CD;$$

the perpendicular falling on the base,

$$BD = BC - DC.$$

Squaring both members,

$$\overline{BD}^2 = \overline{BC}^2 + \overline{DC}^2 - 2BC \times DC,$$

and by adding $\overline{AD}^2$ to each member,

$$\overline{BD}^2 + \overline{AD}^2 = \overline{BC}^2 + \overline{DC}^2 + \overline{AD}^2 - 2BC \times DC;$$

and by Theorem 8,

$$\overline{AB}^2 = \overline{BC}^2 + \overline{AC}^2 - 2BC \times DC.$$

The same process will give the same result, when the perpendicular falls upon the base produced.

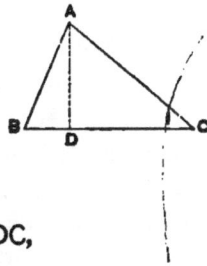

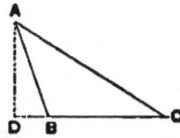

THEOREM X.

In an obtuse-angled triangle, the square of the side opposite the obtuse angle is equivalent to the sum of the squares of the two other sides, plus twice the rectangle of the base and the distance of the obtuse angle from the foot of the perpendicular let fall from the vertical angle on the base produced.

Enunciation,

$$\overline{AC}^2 = \overline{AB}^2 + \overline{BC}^2 + 2BC \times BD,$$
$$CD = BC + BD;$$

by squaring,

$$\overline{CD}^2 = \overline{BC}^2 + \overline{BD}^2 + 2BC \times BD;$$

adding AD² to each side,

$$\overline{CD}^2 + \overline{AD}^2 = \overline{BC}^2 + \overline{BD}^2 + \overline{AD}^2 + 2BC \times BD.$$

Theorem 8. $$\overline{AC}^2 = \overline{BC}^2 + \overline{AB}^2 + 2BC \times BD.$$

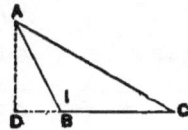

THEOREM XI.

If from the vertex of any angle of a triangle, a line be drawn to the middle point of the opposite side, then twice the square of the bisecting line, plus twice the square of half the bisected side, will be equal to the sum of the squares of the two other sides.

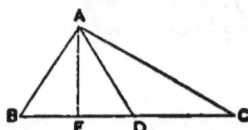

From the vertex A of the triangle ABC draw AD to the middle point of BC; then will

$$2\overline{AD}^2 + 2\overline{BD}^2 = \overline{AB}^2 + \overline{AC}^2.$$

In the triangle ADC, the side AC is opposite the obtuse angle ADC.

$$\therefore \quad \overline{AC}^2 = \overline{AD}^2 + \overline{DC}^2 + 2DC \times DE. \tag{1}$$

And in the triangle ABD the side AB is opposite the acute angle ADB.

$$\therefore \quad \overline{AB}^2 = \overline{AD}^2 + \overline{BD}^2 - 2BD \times DE. \tag{2}$$

By adding equations (1) and (2), and observing that BD = DC,

$$\overline{AB}^2 + \overline{AC}^2 = 2\overline{AD}^2 + 2\overline{BD}^2.$$

Cor.—The sum of the squares of all the sides of a parallelogram, is equivalent to the sum of the squares of the diagonals.

Since the diagonals mutually bisect each other,

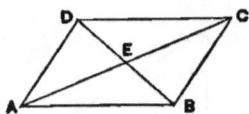

$$\overline{DC}^2 + \overline{BC}^2 = 2\overline{CE}^2 + 2\overline{DE}^2,$$
$$\overline{AB}^2 + \overline{AD}^2 = 2\overline{AE}^2 + 2\overline{DE}^2.$$

By addition, $\overline{AB}^2 + \overline{DC}^2 + \overline{AD}^2 + \overline{BC}^2 = 4\overline{AE}^2 + 4\overline{DE}^2$
(Th. 8, Cor. 2.) $\qquad\qquad\qquad = \overline{AC}^2 + \overline{BD}^2.$

THEOREM XII.

If a line be drawn parallel to one of the sides of a triangle cutting the other sides, it will divide them proportionally.

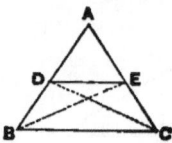

Draw DE parallel to BC, and draw BE and DC; then the triangles DEB and DEC have the same base DE and the same altitude, as both their vertices are in the line BC, parallel to DE; hence, they are equivalent.

The triangles ADE and BDE having the same altitude, as they have a common vertex E, are to each other as their bases; hence,

ADE : BDE :: AD : BD.

The triangles ADE and DEC have a common vertex D; hence,

$$ADE : DEC :: AE : EC;$$

but triangle DEB = triangle DEC, and the two proportions have an equal ratio;

$$\therefore AD : BD :: AE : EC,$$

and by composition,

COR. 1. $AD+BD : BD :: AE+EC : EC;$

that is, $AB : BD :: AC : EC,$

and $AD+BD : AD :: AE+EC : AE,$

that is, $AB : AD :: AC : AE.$

COR. 2.—If any number of lines be drawn parallel to a side of a triangle, the other sides will be cut proportionally.

COR. 3.—If any number of lines be cut by the parallels, they will be cut proportionally.

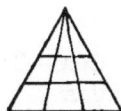

THEOREM XIII.

A line which bisects an angle of a triangle, divides the opposite side into segments proportional to the adjacent sides.

Let AD bisect the angle A; then

$$BD : DC :: AB : AC.   (1)$$

From C draw a line parallel to DA, intersecting BA produced in E; then

$$BD : DC :: AB : AE.   (2)$$

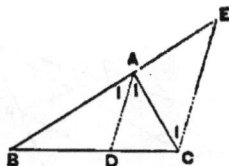

The angle CAD = ACE, alternate; CAD = BAD, bisected; and BAD = BEC, corresponding; $\therefore$ ACE = AEC, and the triangle AEC is isosceles; hence, side AE = AC, and (2),

$$BD : DC :: AB : AC = AE.$$

THEOREM XIV.

Triangles which are mutually equiangular have the sides opposite the equal angles respectively proportional, and hence the triangles are called similar.

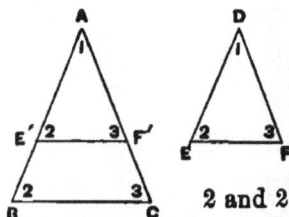

AB : DE :: AC : DF :: BC : EF.

The triangles ABC and DEF are mutually equiangular; that is, A=D, E = B, and F = C. Place DEF on ABC, the point D on A, the side DE on AE', and DF will fall on AC, since angle D = angle A. Since the angles 2 and 2 are equal, E'F' is parallel to BC; and

AB : AC :: AE' : AF',

or,   AB : AC :: DE : DF;

and by alternation,   AB : DE :: AC : DF.

By placing F on C, we obtain the proportion,

AC : DF :: BC : EF.

Cor.—Two triangles having two angles respectively equal are similar.

Rem.—The sides opposite the equal angles are called *homologous.*

THEOREM XV.

The figure formed by joining the middle points of any quadrilateral by straight lines is a parallelogram.

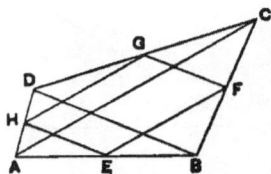

Let ABCD be any quadrilateral. Join the middle points of the sides, and draw the diagonals.

BE : BF :: EA : FC
AE : AH :: EB : HD
DH : DG :: HA : GC
CF : CG :: FB : GD.

It follows that GF is parallel to DB, and HE is also parallel to DB. ∴ GF and HE are parallel; so also EF and HG are parallel.

THEOREM XVI.

Two triangles which have their sides respectively pro-portional are similar.

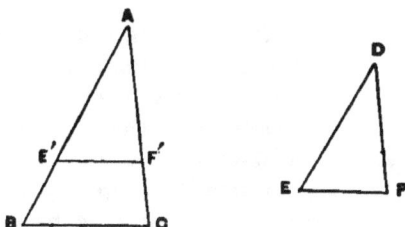

Since the sides are respectively proportional, then

$$AB : DE :: AC : DF :: BC : EF \qquad (1)$$

Make AE′ = DE, and draw E′F′ parallel to BC; then will

$$AB : AE' :: AC : AF' :: BC : E'F'; \qquad (2)$$

but   AE′ = DE.

The proportions (1) and (2) have an equal ratio, hence the other ratios must be the same; hence

$$AF' = DF$$

and   E′F′ = EF;

∴   the triangle AE′F′ = DEF;

but the triangles AE′F′ and ABC are equiangular, as E′F′ is parallel to BC; and as they are equiangular they are similar.

Cor. 1.—If two triangles have each an equal angle included by proportional sides, they are similar.

Cor. 2.—Two triangles which have their sides respectively parallel or perpendicular to each other are similar.

THEOREM XVII.

Regular polygons of the same number of sides are similar figures.

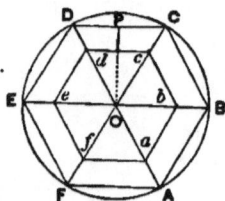

Construct a regular polygon, as in Problem VII, Book II, and let a smaller one be placed upon it, the angles being the same in both polygons; they will also be the same in the isosceles triangles; consequently the sides AB and *ab*, also BC and *bc*, etc., will be parallel; hence the proportions,

$$AB : ab :: R : r,$$

$$BC : bc :: R : r, \quad etc.$$

∴ the triangles are similar; and as each polygon is composed of an equal number of similar triangles, the polygons are similar.

COR. 1.—It is evident that a circumference may be inscribed in the polygon, as the perpendicular OP, which is called the *apothegm* of the polygon is the distance from the center to each side.

COR. 2.—As the equal sides of the isosceles triangles become radii of the circumscribed circle, a circle may be passed through all the vertices.

COR. 3.—Circles are similar figures.

DEF.—Two polygons which are mutually equiangular and have their corresponding sides proportional, are similar.

THEOREM XVIII.

In a right-angled triangle, if a line be drawn from the right angle perpendicular to the hypothenuse, it will divide the given triangle into two triangles, similar to the given triangle and similar to each other.

Let ABC be right-angled at C, and CD perpendicular to the hypothenuse AB. The triangles ABC and ADC have the angle A common, and each has a right angle; they are therefore similar. And for the same reason ABC and BCD are similar; consequently, ADC and BCD are similar.

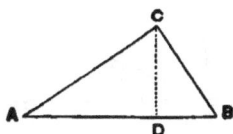

Cor. 1.   Since ABC and ADC are similar,

$$AB : AC :: AC : AD,$$

$$\therefore \quad AB \times AD = \overline{AC}^2. \tag{1}$$

Since ABC and BCD are similar,

$$AB : BC :: BC : BD,$$

$$\therefore \quad AB \times BD = \overline{BC}^2. \tag{2}$$

Since ADC and BCD are similar,

$$AD : DC :: DC : BD; \tag{3}$$

whence,

$$\overline{DC}^2 = AD \times BD.$$

Adding (1) and (2), $\overline{AB}^2 = \overline{AC}^2 + \overline{BC}^2.$

From proportions (1) and (2), the result is that the square of the hypothenuse is equivalent to the sum of the squares of the other sides ; and from (3), that the perpendicular is a mean proportional between the segments of the hypothenuse.

Cor. 2.   If from any point in the circumference of a circle a perpendicular be drawn to the diameter, it will be a mean proportional between the segments of the diameter.

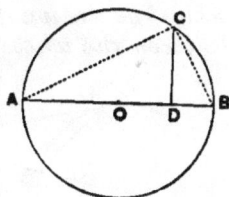

Let C be the point in the circumference from which is drawn the perpendicular CD to the diameter AB; by drawing AC and CB, ABC becomes right-angled at C; hence, $\overline{CD}^2 = AD \times BD.$ (3)

3

THEOREM XIX.

If two chords intersect each other in a circle, their segments are reciprocally proportional.

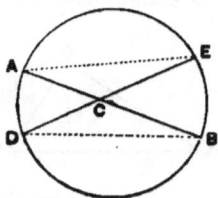

The triangles ACE and BCD are similar; the angle E = angle B, and A = D, respectively measured by ½ the same arc; hence, AC : DC :: CE : BC.

Cor. AC × BC = DC × CE, the product of the segments of the one chord equal to the product of the segments of the other chord.

THEOREM XX.

If from a point without a circle, two secants be drawn terminating in the concave arc, the whole secants will be reciprocally proportional to their external segments.

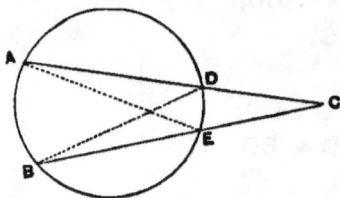

In the similar triangles ACE and BCD,

AC : BC :: CE : DC.

Cor. AC × DC = BC × CE.

THEOREM XXI.

If from a point without the circle a tangent and a secant be drawn, the tangent will be a mean proportional between the whole secant and its external segment.

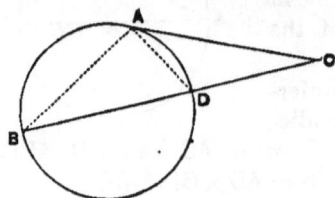

The similar triangles ABC and ACD give the following proportion:

CB : AC :: AC : CD;

hence, CB × CD = AC².

THEOREM XXII.

If a line be drawn bisecting an angle of a triangle and intersecting the opposite side, the rectangle of the sides about the bisected angle equals the rectangle of the segments of the third side plus the square of the bisecting line.

Circumscribe a circle about the given triangle ABC, and bisect the angle C and extend the bisecting line to the circumference of the circle and draw BE. The triangles ADC and BCE are similar; hence,

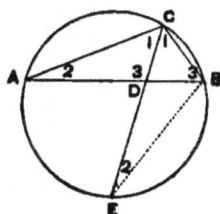

$$AC : CE :: CD : BC;$$

$$\therefore \quad AC \times BC = CE \times CD;$$

but

$$CE = CD + DE,$$

and

$$(CD + DE) \times CD = DE \times CD + \overline{CD}^2,$$

and

$$DE \times DC = AD \times DB.$$

$$\therefore \quad AC \times BC = AD \times BD + \overline{CD}^2.$$

THEOREM XXIII.

Two triangles, having an angle in each equal, are to each other as the rectangles of the sides containing the equal angles.

Let the triangles ABC and ADE have the angle A common; then will ABC : ADE :: AB × AC : AD × AE. Draw BE; then the triangles ABE and ADE have the same altitude, and hence ABE : ADE :: AB : AD; the triangles ABC and ABE have the same altitude, ABC : ABE :: AC : AE. Multiplying the proportions and observing that ABE is common to antecedent and consequent,

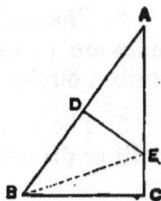

$$ABC : ADE :: AB \times AC : AD \times AE.$$

THEOREM XXIV.

Similar triangles are to each other as the squares of their homologous sides.

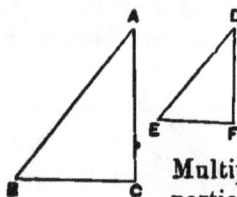

Let ABC and DEF be similar triangles; angle A = angle D,

$$ABC : DEF :: AB \times AC : DE \times DF. \quad (1)$$
$$AB : DE :: AC : DF. \quad\quad\quad (2)$$

Multiplying this proportion by the identical proportion,

$$AC : DF :: AC : DF, \quad\quad\quad\quad (3)$$
$$AB \times AC : DE \times DF :: \overline{AC}^2 : \overline{DF}^2. \quad (4)$$

Since the 1st and 4th have equal ratios,

$$ABC : DEF :: \overline{AC}^2 : \overline{DF}^2.$$

and as the homologous sides are proportional, so also the triangles are to each other as

$$\overline{AB}^2 : \overline{DE}^2 \quad \text{and} \quad \overline{BC}^2 : \overline{EF}^2.$$

Cor. 1.—The areas of regular polygons are to each other as the squares of the radii of the inscribed or circumscribed circle.

Cor. 2.—The areas of circles are to each other as the squares of their radii, or the squares of the diameters.

GENERAL COROLLARIES.

1. The perimeters of similar polygons are to each other as their homologous sides, or as their corresponding diagonals.

2. The perimeters of regular polygons of the same number of sides are to each other as the radii of the inscribed or circumscribed circles.

3. The circumferences of circles are to each other as their radii or diameters.

PROBLEM I.

To divide a given line into five equal parts.

Let AB be the given line. From A draw an indefinite line AH, making any angle with AB, and on it lay off the same distance five times. Join the last point C with B, and from each point draw lines parallel to CB; then AB will be divided into five equal parts. (Th. 12, Cor. 2.)

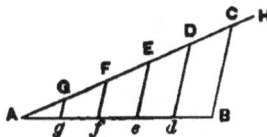

PROBLEM II.

To divide a given line into parts proportional to several given lines.

Let AB be the given line. Draw AH an indefinite line, and on it lay off the several given lines *a*, *b*, *c*, *d*, and join the last point with B, and from each point draw lines parallel to this line; then will the parts A*a'*, *a'b'*, *b'c'*, and *c'*B be proportional to the given lines *a*, *b*, *c*, and *d*.

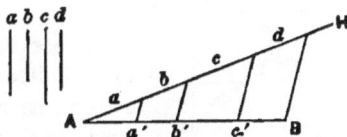

PROBLEM III.

To find a fourth proportional to three given lines.

From any point, as A, draw two indefinite lines AH and AY; on AY lay off *a*, and on AH lay off *b* and join *ab*; on AY lay off *c* and draw *cd* parallel to *ab*; *bd* will be a fourth proportional.

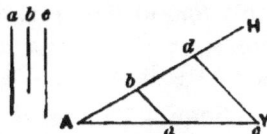

PROBLEM IV.

To construct a mean proportional to two given lines.

On an indefinite line lay off AB and BC, equal respectively to the given lines. On AC describe a semi-circumference, and at B erect the perpendicular BD, which will be a mean proportional between AB and BC. (Th. 18, Cor. 2.)

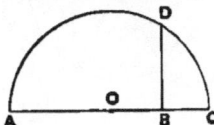

PROBLEM V.

To construct a triangle equivalent to a given polygon.

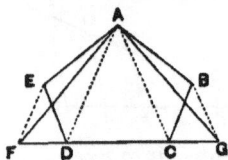

Let ABCDE be the given polygon. From A draw the diagonals AD and AC; then from E and B draw EF and BG, respectively parallel to the diagonals AD and AC, intersecting the base produced; then AFG will be the required triangle.

PROBLEM VI.

To inscribe a square in a circle and circumscribe a square about a circle.

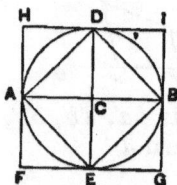

Draw two diameters at right angles and join their extremities, and we have an inscribed square, as each side is a chord of ninety degrees, and each angle is measured by one-half a semi-circumference. At each extremity of the perpendicular diameters draw tangents to the circumference and we have the circumscribed square.

Cor. 1.—The circumscribed square has double the area of the inscribed, as it has eight equal triangles, whilst the inscribed has only four of the equal triangles; hence, area of cir. sq. : area of ins. sq. :: 2 : 1, and side of cir. sq. : side of ins. sq. :: $\sqrt{2}$: 1; same result as in Th. 8, Cor 2.

Scho.—The side of the circumscribed square is the same as the diagonal of the inscribed.

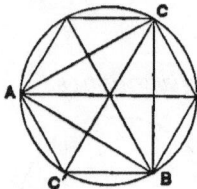

Cor. 2.—In the triangle CBC', right-angled at B, we have

$$\overline{CC'}^2 - \overline{C'B}^2 = \overline{CB}^2.$$

$$CC' = 2, \quad \text{and} \quad C'B = 1.$$

$$\therefore \quad 4 - 1 = \overline{CB}^2,$$

$$\text{and} \quad \overline{CB}^2 = 3,$$

$$CB = \sqrt{3}.$$

That is, *The side of an equilateral triangle : Radius ::* $\sqrt{3}$: 1.

PROBLEM VII.

To divide a given line into extreme and mean ratio; that is, into two such parts that the greater part shall be a mean proportional between the whole line and the less part.

Let AB be the given line. At B erect a perpendicular BC equal to ½AB; and with C as a center and radius CB, describe a circumference. From A draw AF through the center and terminating in the concave arc, and with A as a center and AD as radius, draw the arc DE, making AE equal to AD; then DF = AB, and (Theorem 21) AF : AB :: AB : AD, by inversion AB : AF :: AD : AB, and by division AB : AF — AB :: AD : AB — AD; that is, AB : AD :: AD : EB, or AB : AE :: AE : EB.

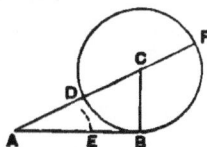

PROBLEM VIII.

To construct a square equivalent to a given triangle.

A mean proportional between the base and half the altitude of the triangle will be a side of the square.

Let B = Base and A = ½ Altitude. DC is a mean proportional between base and one-half altitude.

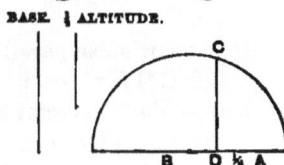

PROBLEM IX.

To construct a square equivalent to two given squares.

Construct a right angle. On one of the sides of the angle lay off a distance equal to a side of one of the squares; and on the other side of the angle a distance equal to a side of the other square, and draw the hypothenuse; it will be a side of the required square.

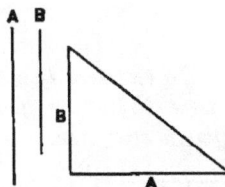

REM.—By this principle the side of a square equivalent to any number of squares may be found.

COR.—By making the longer side the hypothenuse, the third side will be the side of a square equal to the difference of two squares.

REM. 1.—If similar polygons be constructed on the three sides of a right-angled triangle, the given sides being homologous, the polygon constructed on the hypothenuse will be equivalent to the sum of the two others.

REM. 2.—To construct a square equivalent to a given polygon, reduce the polygon to an equivalent triangle and find a mean proportional between the base and half the altitude of the triangle.

PROBLEM X.

To construct a polygon, similar to a given polygon, on a given side homologous to one of the sides of the given polygon.

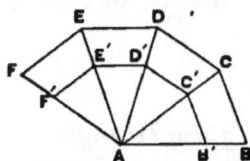

Let ABCDEF be the given polygon and AB' a side of the required polygon homologous to AB. Lay off AB' on AB, and from A draw all the diagonals. Draw B'C' parallel to BC to the first diagonal; then from one diagonal to another draw sides parallel to the opposite side of the given polygon. AB'C'D'E'F' will be the required polygon. (Th. 17, Cor. 4.)

COR.—To construct a regular polygon, having one of the sides given: First construct a regular polygon of the proper number of sides; then find a fourth proportional to the side of the constructed polygon, the side of the required polygon, and the radius of the circumscribed circle of the constructed polygon; the fourth proportional will be the radius of the circumscribed circle of the required polygon.

PROBLEM XI.

To extract the square root of a quantity, or, what is the same thing, to find the side of a square equivalent to a given surface.

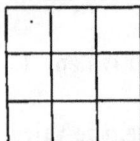

The surface of a square is found by squaring a side; thus, $3 \times 3 = 9$, that is, 3 in length and 3 in breadth. 9 is the surface of which we wish to find a side of a square equivalent; and, as $3 \times 3 = 9$, it is evident that 3 is the square root of 9; so also 4 of 16, 5 of 25, 6 of 36, etc.; but when the number is large it is not so easily found.

Let us take an algebraic binomial, as $(a+b)^2$ $= a^2 + 2ab + b^2$, and exhibit it geometrically.

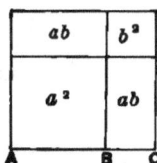

The divisors must be such as to render the quotient a root $(a+b)$.

$$a \,)\; a^2 + 2ab + b^2 \,(\, a + b$$
$$\quad a^2$$
$$2a + b \,)\; \overline{+\, 2ab + b^2}$$
$$\qquad\qquad +\, 2ab + b^2$$

Next take a trinomial; as

$$(a+b+c)^2 =$$
$$a \,)\; a^2 + 2ab + b^2 + 2ac + 2bc + c^2 \,(\, a + b + c$$
$$\quad a^2$$
$$2a + b \,)\; 2ab + b^2$$
$$\qquad\quad 2ab + b^2$$
$$2a + 2b + c \,)\; +\, 2ac + 2bc + c^2$$
$$\qquad\qquad\qquad +\, 2ac + 2bc + c^2$$

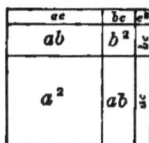

COR. 1.—The first term of the roots is obtained the same as that of finding the root of a monomial; the divisor and root are the same, as the first surface a^2 is a square; after that we have rectangles, the breadth of which is the root, and our divisor must be the whole length of the rectangles.

COR. 2.—Each successive divisor is double the root already found, plus the next term of the root.

REM.—In Arithmetic we pursue the same course, except that the squares are not so entire and separate as in Algebra and Geometry; hence, in general, we take the nearest root, the largest figure of which the square is less than the given number, and we point off the figures in periods of two each, beginning at the unit's place. The reason of pointing off in periods is shown by the increase of the numbers in squares; thus,

1	11	9	99
1	11	9	99
1	121	81	9801

The increase of one figure in the side makes two in the surface; it will always be this, and never more or less, as is shown by taking the smallest and the largest digits.

PROBLEM XII.

To find the circumference of a circle whose radius is unity.

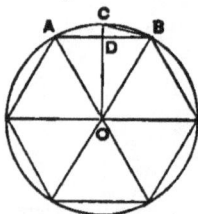

With 1 as a radius describe a circumference, and inscribe in it a regular hexagon, each side of which will be unity. Take any side, as AB, and bisect it in D and its arc in C, and draw the chord CB, which will be the side of a regular polygon of double the number of sides. The triangle ODB is right-angled at D; hence,

$$OD = \sqrt{\overline{OB}^2 - \overline{DB}^2} = \sqrt{1 - \tfrac{1}{4}} = \tfrac{1}{2}\sqrt{3},$$

and $\quad CD = 1 - \tfrac{1}{2}\sqrt{3}, \quad$ and $\quad CB = \sqrt{\tfrac{1}{4} + (1 - \tfrac{1}{2}\sqrt{3})^2}.$

Let C represent a side of the first polygon, and *c* a side of the polygon of double the number of sides ; in each successive computation, after *c* is found, make it C in the next, and continue this process until the difference between C and *c* has no appreciable value ; then this value of C multiplied by the number of sides will give 6.2832, which is the approximate length of the circumference when the radius is 1 and the diameter 2.

When the diameter is 1, the circumference is 3.1416, which number is represented by π; hence, πd or $2\pi r$ represents circumference.

PROBLEM XIII.

The area of a regular polygon is equal to its perimeter multiplied by one-half the radius of the inscribed circle.

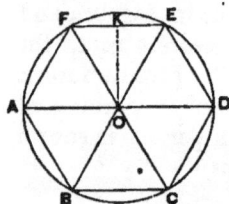

Let ABCDEF be a regular inscribed hexagonal polygon, and OK the radius of the inscribed circle. The polygon is composed of six triangles, each having a side of the polygon for its base and OK for its altitude ; hence the area of the polygon is its perimeter multiplied by $\tfrac{1}{2}$OK ; that is, $\tfrac{1}{2}$ the radius of the inscribed circle.

Cor.—When the number of sides of the polygon is indefinitely increased, it becomes a circle, and the radius of the inscribed circle, which has been increasing as the number of sides increased, is now the radius of the circle, and the perimeter of the polygon is the circumference of the circle ; hence, the area of a circle is equal to its circumference multiplied by one-half its radius.

BOOK V.

THEOREM I.

When the distance between the centers of two circles is greater than the sum of their radii, they are external, and the straight line joining their centers will be the shortest distance between the center of either circle and the circumference of the other; and if this line be extended to the concave arcs of both circles, it will be greater than between any two other points in the circumference.

Let C and C′ be the centers of two circles external to each other; C′B is the shortest distance from C′ to any point in the

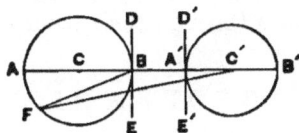

circumference C; for, let the tangents DE and D′E′ be drawn at B and A′, they will be perpendicular to C′C; and as a perpendicular is the shortest distance from a point to a line, C′B is the shortest distance from C′ to the tangent DE, and any other line from C′ to the circumference is oblique to the line DE, and must go beyond it before it can reach the circumference.

2. C′A is longer than any other line drawn from C′ to the circumference C, as CF. Draw the chord BF; BF is less than AB a diameter; hence,

$$C'B + BF < C'A,$$

and $$C'F < C'B + BF;$$

much more then is $$C'A > C'F.$$

THEOREM II.

When the distance between the centers is equal to the sum of the radii, they are tangents externally, and the straight line joining their centers passes through the point of tangency.

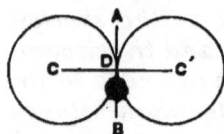

They must touch on the line joining their centers, as CD + DC′ = CC′; let D be this point, and through D draw AB perpendicular to CC′; it will be a common tangent to both circles as it is perpendicular to each radius at its extremity.

Or, CC′ is the shortest distance between the centers, and C′D is the shortest distance from C to AB; CD is also the shortest distance from C to AB; therefore, the line between the centers of the two circles passes through their point of tangency.

THEOREM III.

When the distance between the centers is less than the sum and greater than the difference of the radii, they intersect each other, and the line joining their centers is perpendicular to their common chord at its middle point.

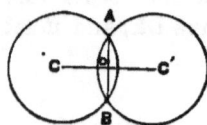

CC′ is the shortest distance between C and C′; hence, CD and C′D must both be perpendicular to AB at its middle point, and CC′ must be a straight line.

THEOREM IV.

When the distance between the centers is equal to the difference of the radii, the smaller circle is tangent internally to the larger one, and the line joining their centers extended passes through the point of tangency.

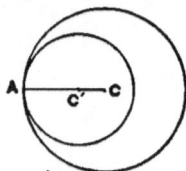

C′A is the shortest distance from the center C′ to the circumference C; therefore, A is the point of tangency.

THEOREM V.

When the distance between the centers is less than the difference of the radii, the smaller is wholly within the larger, and the nearest and the farthest points in the circumference of the one circle, from the center of the other circle, is in the extensions of the line joining their centers.

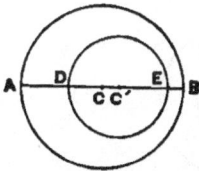

C'B is the nearest distance and C'A is the farthest in the circumference C from the center C'. (Th. 1.)

COR.—When they are concentric, their circumferences are parallel.

GEN. COR.—The line joining the centers passes through the points of tangency, the middle point of the common chord, the nearest and the farthest point of the circumference.

THEOREM VI.

If on one side of a given polygon another polygon be constructed within the given polygon, the perimeter of the interior polygon will be less than that of the given polygon.

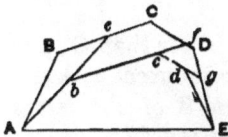

Produce each side of the interior polygon until it meets a side of the exterior. Then A*e* is less than AB*e*,

$$bf < be\text{C}f, \quad cg < cf\,\text{D}g, \quad \text{and} \quad d\text{E} < dg\text{E}.$$

$$\therefore \ \text{A}e\text{CDE} < \text{ABCDE}; \quad \text{A}bf\,\text{DE} < \text{A}e\text{CDE};$$

$$\text{A}bcg\text{E} < \text{A}bf\,\text{DE}; \quad \text{and} \quad \text{A}bcd\text{E} < \text{A}bcg\text{E};$$

much more is A*bcd*E less than ABCDE.

COR.—If from any point within a triangle lines be drawn to the extremities of either side, the sum of these lines will be less than the sum of the two other sides.

THEOREM VII.

If two circles whose radii are unequal intersect each other, the middle point of their common chord will be nearer the arc of the large circle than that of the smaller one.

Let C and C′ be the centers of the two circles; then will DF be less than DE. For place the two circles so that the smaller one shall be tangent internally to the larger one.

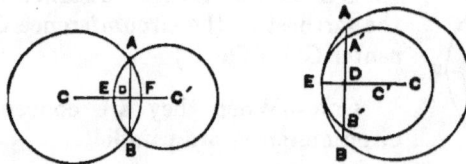

Draw AB a chord to the larger circle, A′B′ will be a chord of the smaller one at the same distance from the arcs of the circles; the chord AB is longer than the chord A′B′, and in order that A′B′ become equal to AB it must be put nearer the center of the circle and farther from the arc; therefore, DF is less than DE.

THEOREM VIII

If the circumferences of two unequal circles intersect each other, the arc of the larger circle is less than that of the smaller one.

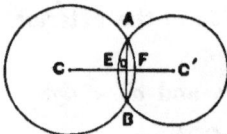

Let the circle C be larger than C′, and let their circumferences intersect at A and B; then will the arc AFB of the larger circle be less than the arc AEB of the smaller one.

Since DF is less than DE (Th. 7), the arc AFB may be revolved on AB as an axis, the point F will fall upon ED, between D and E, and the arc AFB will be wholly within the arc AEB, and is therefore less. (Th. 6.)

Rem.—An arc may be regarded as a portion of the perimeter of a polygon.

BOOK VI.

PLANES IN DIFFERENT POSITIONS.

.

_____

DEFINITIONS.

1. A straight line is **perpendicular** to a plane, when it is perpendicular to every line passing through its foot in that plane.

2. Two planes are parallel when they are everywhere equally distant, and consequently will never meet.

3. Three points not in the same straight line determine the **position** of a plane.

4. The **intersection of two planes** is a straight line, for two planes cannot have three points common which are not in the same straight line; for, if they have, they form but one plane.

5. A Diedral Angle is the divergence of two planes from their common intersection, and is measured by the plane angle formed by two lines, one in each plane, perpendicular to the common intersection, at any point in this line.

REM.—Theorems regarding several planes in different positions have no distinct principles; but special attention is required to the position of the planes, as a correct figure of two or more planes cannot be drawn upon one plane, as the blackboard.

The few theorems given on this subject are thought sufficient to illustrate this peculiarity.

THEOREM I.

If from a point without a plane, a perpendicular be drawn to the plane and oblique lines to different points of the plane,

1st. Any oblique line will be longer than the perpendicular.

2d. Oblique lines drawn to points equally distant from the foot of the perpendicular will be equal.

3d. Oblique lines unequally distant from the foot of the perpendicular, the one farther distant will be the longer.

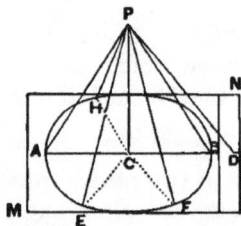

Let P be a point without the plane MN, PC a perpendicular to the plane, and A and B two points in the plane MN equally distant from C the foot of the perpendicular; and D a point farther distant from C than A and B.

1st. Let all the points A, C, B, and D be in the same straight line. Suppose a plane passed through them and the point P; then will all the lines PA, PC, PB, and PD be in the plane PAD; and AD, the line of intersection of the two planes, will be a base line for the figure ADP. Since PC is perpendicular to the plane MN, it is perpendicular to AD. (Def. 1.) PA and PB are oblique lines drawn from P to points equally distant from the foot of PC; hence, PA = PB. (Bk. 2, Th. 6, part 2.) And as D is farther distant from C, PD is longer than PA or PB. (Part 3, same Prop.)

2d. Other oblique lines, as PE, PF, and PH, may be drawn to different points of MN, at the same distance from C as A and B, and the planes PCE, PCF, and PCH passed; the triangles PCE, PCF, and PCH will be right-angled triangles, equal to PCA and PCD; hence, PE = PF = PA = PB.

REM.—If a circumference be drawn with C as a center and a radius equal to CA, it will pass through all the points A, E, F, B, and H, and the point D will be without the circumference.

THEOREM II.

A line which is perpendicular to two lines of a plane, at their intersection, is perpendicular to any other line of the plane passing through this intersection and therefore perpendicular to the plane.

Let PC be perpendicular to AB and DE at C, the point of intersection of the two lines in the plane MN. With C as a center and any radius describe a circumference cutting the two lines in A, B, D, and E, and draw PA, PB, PD, and PE. Through C draw any other line as FG, terminating in the circumference. PC will also be perpendicular to FG; as F and G are equally distant from C, PF is equal to PG; hence the line PC has two points P and C equally distant from F and G, the extremities of the line FG; hence PC is perpendicular to FG, any line of the plane MN, and is therefore perpendicular to every line in the plane; consequently perpendicular to the plane. (Def. 1.)

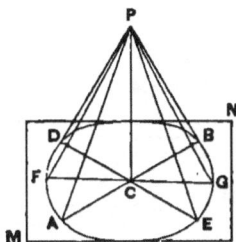

Cor.—The two sides of an angle, two parallel lines, or three points not in the same straight line, determine the position of a plane.

THEOREM III.

If from the foot of a perpendicular to a plane a line be drawn at right angles to any line of that plane, and the point of intersection of these two lines be joined with any point in the perpendicular, the last line will be perpendicular to the line in the plane to which the line was drawn at right angles.

From P, the foot of the perpendicular AP, draw PB perpendicular to HY, any line in the plane MN. Make BC and BD equal, and draw PC and PD, which will also be equal; and then draw AC and AD, which will also be equal. (Bk. 2, Prob. 6, part 2.) And since AB has two points, A and B, equally distant from D and C, AB is perpendicular to DC. (Bk. 1, Th. 9.)

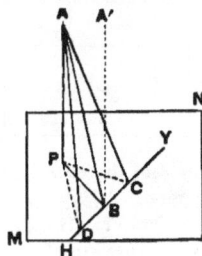

Cor. 1.—Since BA is perpendicular to DC, when drawn from any point in the line PA or PA produced indefinitely; and as in every position BA lies in the plane APB, it is also perpendicular to DC when it becomes parallel to PA and is also perpendicular to PB, and hence perpendicular to the plane MN; consequently, if one of two parallels is perpendicular to a plane, the other is also perpendicular to the same plane.

Cor. 2.—Two lines perpendicular to the same plane are parallel.

THEOREM IV.

If two parallel planes are intersected by a third plane, the lines of intersection will be parallel.

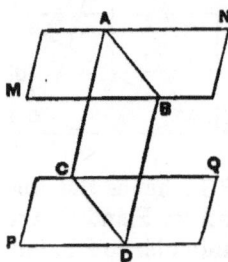

Let the parallel planes MN and PQ be intersected by the plane ABCD; then will the lines of intersection AB and CD be parallel.

Make AB and CB of the same length, and draw AC and BD; the line AB lies in the plane MN, and CD in the plane PQ; as the planes are parallel, they will never meet; hence, AB and CD will never meet; but AB and CD also lie in the same plane ABCD, and are therefore parallel.

Cor. 1.—The figure ABCD is a parallelogram.

Cor. 2.—Parallel lines intercepted between parallel planes are parallel.

Rem.—A plane may always be made to pass through parallel lines, as all parallels have the same direction.

THEOREM V.

If in parallel planes, angles are formed whose sides respectively take the same direction, the angles will be equal.

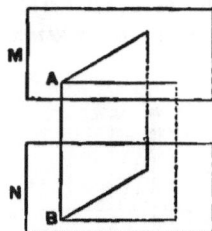

In the planes M and N, let A and B be angles whose sides respectively take the same direction. Since the sides of the angles respectively take the same direction, if the one angle is placed on the other, they will coincide, and consequently are equal.

Cor. 1.—If the vertices A and B are joined by a straight line, and planes passed through this line and the corresponding sides of the two angles, a diedral angle will be formed.

Cor. 2.—If a diedral angle is cut by parallel planes, the plane angles formed are equal.

BOOK VII.

DEFINITIONS.

1. A Prism is a solid, two of whose faces are equal parallel polygons, which are termed the lower and upper bases, whilst the other faces are parallelograms which form the convex surface.

REM.—The bases may be polygons of any number of sides.

2. A Right Prism has its lateral edges perpendicular to its bases.

3. An Oblique Prism has its lateral edges oblique to its bases.

4. The **Altitude** of a prism is the perpendicular distance between its bases.

5. A Regular Prism is a right prism having regular polygons for its bases.

6. A Parallelopipedon is a prism having its bases and its faces parallelograms, the opposite faces necessarily equal.

7. If the bases and faces are rectangles, it is called a **Rectangular Parallelopipedon.**

8. If they are all equal squares, it is a **Cube.**

REM.—A solid is said to have three dimensions, to which special attention must be given ; thus, the dimensions of a rectangular parallelopipedon are the length and the perpendicular breadth of the base, regarding it as a plane figure, and the perpendicular distance between the two bases.

In a prism of any shape, the dimensions of the base are the same as those of a plane figure, and the altitude is the perpendicular distance between the bases.

THEOREM I.

The sections formed by parallel planes cutting a prism, are equal polygons.

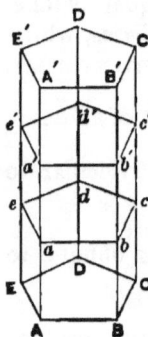

Let *abcde* and *a'b'c'd'e'* be sections formed by parallel planes cutting the prism ; *ab* and *a'b'* are parallel, being the intersection of parallel planes with a third plane (Book 6, Th. 4); and they are equal, as they are parallels between two other parallels; so also for the same reasons are *bc = b'c'*, *cd = c'd'*, etc. ; and since the angles are formed by parallel planes cutting diedral angles, they are respectively equal (Bk. 6, Th. 5, Cor. 2); hence, the sections are equal polygons.

Cor.—If the sections are parallel to the bases, then will they be equal polygons.

THEOREM II.

The lateral surface of a right prism is equal to the product of its perimeter and altitude.

As each face is a rectangle, its surface is equal to the product of its base and altitude; hence, the lateral surface of the prism is the product of the sum of the sides of the base and the common altitude; but the sum of the sides of the base is the perimeter; hence, the lateral surface of a prism is the product of its perimeter and altitude.

Cor. 1.—If the prism be oblique, its faces will be parallelograms instead of rectangles, and is measured accordingly.

Cor. 2.—The convex surface of a cylinder is equal to the product of the circumference of its base and altitude. For a prism having a regular polygon for its base becomes a cylinder, by increasing the number of sides indefinitely, and the perimeter of the polygon, forming its base, becomes the circumference of the base of the cylinder.

Cor. 3.   Convex surface of a cylinder $= 4\pi R^2$,
Surface of the two bases $= 2\pi R^2$,
Entire surface of the cylinder $= \overline{6\pi R^2}$.

THEOREM III.

The volume of a rectangular parallelopipedon is equal to the product of its three dimensions.

The area of the rectangular base is the product of its two dimensions, and for every unit in altitude there will be as many solid units as there are square units in the base; hence, the volume is equal to the product of the three dimensions.

Cor. 1.—A rectangular parallelopipedon can be divided into two equal triangular prisms; and the volume of each is equal to the area of the base multiplied by the altitude.

Cor. 2.—The volume of any right prism is equal to the area of its base multiplied by its altitude; as any prism may be divided into triangular prisms.

Cor. 3.—A right prism having for its base a regular polygon, if the number of sides of the polygon be indefinitely increased, it becomes a cylinder; hence the volume of a cylinder is equal to the area of the base multiplied by its altitude.

Cor. 4.—An oblique prism is measured in the same way; as the area of a rectangle and a parallelogram of the same dimensions is equal; so also is the volume of the oblique prism equal to that of the right prism of the same dimensions.

Scho. 1.—The above results are true if one or more of the dimensions are fractional, as the common denominator of the fractions will be the denomination of the unit of measure.

Scho. 2.—If the dimensions are incommensurable, the unit of measure will be an infinitesimal.

THEOREM IV.

A triangular prism may be divided into three equivalent pyramids.

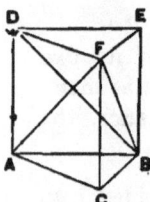

1st. Let ABCDEF be a triangular prism. Pass a plane through the three points, A, B, and F, cutting off the pyramid ABC-F, having the triangle ABC for its base, and the altitude of the prism for its altitude.

2d. Pass a plane through BFD, cutting off the pyramid DEF-B, the upper base for its base, and having the altitude of the prism for its altitude; hence it has the same dimensions as the first pyramid, and is consequently equal to it.

3d. Take away the first pyramid and place the two remaining upon ABDE as a base; their vertices will be in F, and hence will have the same altitude ; their bases are equal, as each is the half of the parallelogram ABED, as the diagonal BD divides it into equal parts.

And as the first and second are equal, and also the second and third, it follows that the first and third are equal; hence they are all equal.

Cor.—The volume of a triangular pyramid is equal to the area of its base, multiplied by one-third its altitude.

THEOREM V.

The volume of the frustum of a triangular pyramid is equivalent to that of three pyramids, two of which shall have for their altitudes the altitude of the frustum, and for their bases respectively the lower and upper bases of the frustum, and the third pyramid shall be a mean proportional between the other two.

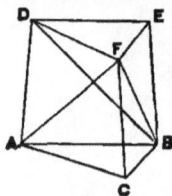

Let ABCDEF be the frustum of a pyramid.

1st. Pass a plane through ABF, cutting off a pyramid, having the lower base of the frustum for its base, and the altitude of the frustum for its altitude. Designate it P.

2d. Draw the diagonal DB and pass a plane through DBF, cutting off a pyramid, having for its base the upper base of the frustum, and for its altitude the altitude of the frustum. Designate this pyramid P'.

Take away the pyramid P and place the remnant on ABED for its base, and being cut by the plane DBF, forms two pyramids, having their vertices in F; hence they have the same altitude: the one P' has DEB for its base; and the third, which designate p, has for its base ABD.

Since P and P' have the same altitudes, they are in proportion to their bases; and as their bases are similar triangles, which are proportional to the squares of their homologous sides, AB and DE are homologous sides; let

$$S = AB,$$

and $$s = DE,$$

then $$P : P' :: S^2 : s^2; \qquad (1)$$

p and P' have the same altitude; hence they are in proportion to the areas of their bases, ABD and DEB, which triangles have the same altitude, and are therefore proportional to their bases, S and s; hence,

$$p : P' :: S : s; \qquad (2)$$

squaring the proportion,

$$p^2 : P'^2 :: S^2 : s^2; \qquad (3)$$

∴ (1) and (3) have an equal ratio; hence,

$$P : P' :: p^2 : P'^2.$$

Multiplying the extremes and means,

$$P \times P'^2 = P' \times p^2;$$

and dividing by P', $$P \times P' = p^2;$$

that is, the third pyramid is a mean proportional between the other two.

THEOREM VI.

The convex surface of a right pyramid is equal to the product of the perimeter of its base and one-half its slant height.

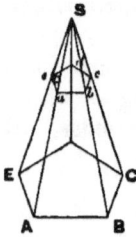

The base of the right pyramid ABCDE is a regular polygon. A perpendicular from the vertex S would fall upon the center of the polygon. The faces will all be equal isosceles triangles; and the surface of each is the product of its base and one-half its altitude; the altitude of each triangle is the slant height of the pyramid; consequently, the entire convex surface is the perimeter of the base multiplied by one-half the slant height.

Cor. 1.—If the number of sides of the regular polygon of the base be indefinitely increased, the polygon becomes a circle, and the pyramid a cone; hence the convex surface of a cone is one-half the product of the circumference of the base and the slant height.

Cor. 2.—If a plane *abcde* cut the pyramid or cone parallel to the base, then the portion between the parallel bases will be a frustum of a pyramid or cone. The sides of the frustum are all trapezoids, each of which is measured by the product of one-half the sum of the parallel bases and altitude; hence, the convex surface of the entire frustum is one-half the product of the sum of the two perimeters and the slant height.

Cor. 3.—The volume of the frustum of a cone is equal to that of three cones, two having for their altitude the altitude of the frustum, and for their bases respectively the lower and upper bases of the frustum, and the third cone a mean proportional between the other two.

BOOK VIII.

DEFINITIONS.

1. A Polyedral Angle is the divergence of three or more planes from the point formed by their common intersection.

2. An angle formed by three planes is called a **Triedral Angle.**

3. The plane angles which form a triedral angle are called **Facial Angles.**

4. A Sphere is a solid every point of whose surface is equally distant from a point within called the centre.

5. The distance from the center to the surface is the **Radius** of the sphere.

THEOREM I.

The sum of any two of the plane angles which form a triedral angle is greater than the third angle.

Let the plane angle ASB be greater than either of the other angles which form the triedral angle S.

In the plane ASB draw AB, making SA = SB, and on the plane angle ASB place the angle BSC; SB on SB, and SC will take the direction of SD, and make SC=SD; then as AB < AC + BC, and BC = BD, and taking BC = BD from each side, AD < AC.

The two triangles ASC and ASD have two sides respectively equal and the third side AC > AD; hence the angle ASC > ASD; therefore the sum of the two plane angles BSC and ASC is greater than the third angle ASB.

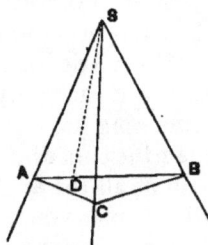

4

THEOREM II.

The sum of all the angles which form a polyedral angle is less than four right angles.

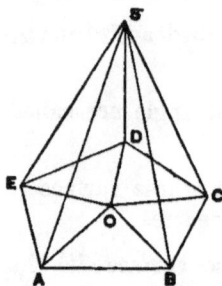

Let S be the vertex of a polyedral angle, and pass a plane cutting the planes and forming the polygon ABCDE.

From any point as O within the polygon, draw lines to the extremities of all its sides; there will be as many triangles as faces forming the polyedral angle.

At each vertex of the polygon there is a triedral angle, formed by one plane angle of the polygon and two in the faces of the polyedral angle S; the one angle in the polygon is less than the sum of the two others (Th. 1); but as the number of the triangles in the polygon is the same as of the plane angles forming the polyedral angle S, hence the sum of all the angles of the triangles in the polygon is equal to the sum of all the angles of the triangles forming the polyedral angle S.

And as the sums of all the angles of the polygon are less than the sums of all the angles on the faces at the points A, B, C. etc., at which triedral angles are formed, on the faces there are two angles at each vertex, whilst in the polygon there is but one, the remaining angles of the triangles of the polygon must be greater than the sum of the angles forming the polyedral angle at S. The sum of all the angles at O is four right angles; hence the sum of all the plane angles forming the polyedral angle S is less than four right angles.

THEOREM III.

The surface generated by the revolution of a regular semi-polygon about the diameter of the circumscribed circle as an axis, is equal to the circumference of the inscribed circle multiplied by this axis.

Let ABCD be one-half of a regular polygon, which being revolved about AD the diameter of the circumscribed circle, as an axis, the surface generated by the perimeter is $2\pi \times Oa \times AD$.

The triangles ABb and CDc will generate equal cones, and the rectangle BbCc will generate a cylinder.

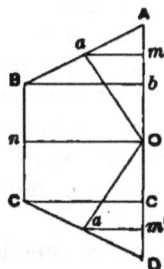

The surface generated by $AB = 2\pi \times am \times AB,$
"　　　"　　　"　　"　$BC = 2\pi \times nO \times BC,$
"　　　"　　,"　　"　$CD = 2\pi \times a'm' \times CD.$

The triangles Oam and ABb are similar; consequently,

$$AB : Oa :: Ab : am ; \quad \therefore \quad AB \times am = Oa \times Ab;$$

hence also, $\qquad a'm' \times CC = Oa' \times cD.$

$$\text{Surface generated by } AB = 2\pi \times Oa \times Ab,$$
$$\text{"} \qquad \text{"} \qquad \text{"} \ BC = 2\pi \times On \times bc,$$
$$\text{"} \qquad \text{"} \qquad \text{"} \ CD = 2\pi \times Oa' \times cD.$$

$Oa = On = Oa'$, each equal to the radius of the inscribed circle. By addition, Whole surface $= 2\pi \times Oa \times AD.$

COR. 1.—When the number of the sides of the semi-polygon is indefinitely increased, it becomes a semicircle; the radius of the inscribed circle becomes equal to that of the circumscribed, and the figure generated a sphere; and its surface $= 2\pi \times r \times 2r = 4\pi r^2$; that is,

The surface of a sphere is equal to the circumference of a great circle multiplied by the diameter.

COR. 2.—The surfaces of spheres are to each other as the squares of their radii; for, let R and R' be the radii of two spheres, their surfaces will be $4\pi R^2$ and $4\pi R'^2$; hence,

$$4\pi R^2 : 4\pi R'^2 :: R^2 : R'^2.$$

THEOREM IV.

The volume of a sphere is equal to the product of its surface and one-third of its radius.

Take a cube, each side say 2 inches, and suppose a sphere inscribed. Consider each face as the base of a pyramid whose vertex is in the center of the inscribed sphere, whose radius is one inch, which is also the altitude of each pyramid.

The volume of each pyramid is equal to the product of its base, and one-third of its altitude; and, the volume of all the pyramids, which equals that of the cube, is the whole surface of the cube multiplied by one-third the radius of the inscribed sphere.

The cube has eight triedral angles; let each of these be cut by a plane tangent to the inscribed sphere, and perpendicular to a straight line drawn from the center of the sphere to the vertex of each triedral angle; thus, a new set of pyramids will be formed, each having the same altitude as the former; the number of bases will be increased, and the sum of all the bases will be the surface of the solid figure remaining.

Continue to pass planes indefinitely, tangents to the inscribed sphere, until the surface of the figure becomes the surface of a sphere, and its volume will be the surface of the sphere multiplied by one-third of the radius; that is,

$$\text{Volume of sphere} = 4\pi R^2 \times \tfrac{1}{3}R = \tfrac{4}{3}\pi R^3;$$

$$\text{Volume of cylinder} = \pi R^2 \times 2R = 2\pi R^3;$$

hence, Sur. of sphere : Sur. of cylinder :: 4 : 6 :: 2 : 3.

Vol. of sphere : Vol. of cylinder :: $\tfrac{4}{3}$: 2 :: 4 : 6 :: 2 : 3;

∴ Sur. sphere : sur. cylinder :: vol. sphere : vol. cylinder.

REM.—When a plane touches a sphere at but one point, it is tangent to the sphere, and the plane is perpendicular to the radius drawn to this point.

THEOREM V.

Every section of a sphere made by a plane is a circle.

Let ACBD be a section of a sphere whose center is O; draw OE perpendicular to the section; OA, OB, OC, and OD, and any other line from O to the intersection of the plane and the surface of the sphere, will all be equal, as they are radii of the sphere; and as OE is perpendicular to the section, it will pass through the middle point of AB or any other line passing through E and terminating in the surface of the sphere; hence, ACBD is a circle and E is its center.

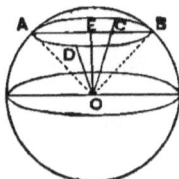

REM.—If the plane pass through the center of the sphere, the circle formed will be a great circle; if it do not pass through the center it will form a small circle.

COR. 1.—A line drawn from the center of a sphere perpendicular to a small circle passes through its center, or a line perpendicular to a small circle at its center passes through the center of the sphere, and its extremities are poles of the small circle and of every circle whose plane is parallel to that of the small circle.

COR. 2.—The pole of a circle is equally distant from every point in the circumference; in the case of the great circle both poles are equally distant, each being a quadrant's distant; in the case of the small circle, the one is greater the other less distant than a quadrant.

COR. 3.—The farther from the center the less the circle.

THEOREM VI.

A great circle divides the sphere into two equal parts.

The plane of the great circle passes through the center of the sphere, and divides it into two parts, each of which has the great circle for its base, and every point of the convex surface of each part is equally distant from the center of their common bases; hence, the two parts must coincide and consequently are equal.

REM.—Each part is called a hemisphere.

THEOREM VII.

The intersection of two great circles is a diameter of the sphere.

The intersection of two planes is a straight line, and as each plane of a great circle passes through the center of the sphere, the center is one point of their intersection; hence their intersection is a straight line passing through the center of the sphere. This straight line is a diameter.

Cor.—The intersection of the circumferences on the surface of the sphere will be the extremities of the diameter, 180° distant; hence they bisect each other.

Def. 1.—The portion of the surface of a sphere included between two semi-circumferences of great circles is called a **Lune.**

Def. 2.—A **Spherical Polygon** is a portion of the surface of a sphere bounded by three or more arcs of great circles. The arcs form the sides of the polygon, each of which is less than a semi-circumference.

Rem.—The sides of a spherical polygon correspond to the facial angles of the polyedral made at the center of the sphere by the planes of the sides of the polygon, and the angles of the polygon correspond to the diedrals of the same polyedral angle.

THEOREM VIII.

The shortest distance between any two points on the surface of a sphere is traced on the arc of a great circle.

The truth of this theorem is evident from Th. 8, Bk. 5. For, when two unequal circumferences intersect on the surface of a sphere, the intercepted arcs hold the same position in regard to each other; as, when a large and a smaller circle intersect on the same plane; from which it is evident that the intercepted arc of the greater circle is less than that of the smaller; and, as on a sphere the circumference of a great circle is the largest that can be described upon it, hence the shortest distance between any two points on the surface of a sphere must be traced on it.

PROBLEM I.

To pass a small circle through any three points on the surface of a sphere, not in the circumference of a great circle.

Let A, B, and C be the three points. Join A and C, and B and C by arcs of great circles. Pass arcs of great circles perpendicularly through the middle points of AC and CB, as in Plane Geometry, and their intersection O will be the pole of a small circle, from which, with the distance between the points of the dividers equal to OA, describe the small circle, which will pass through A, C, and B.

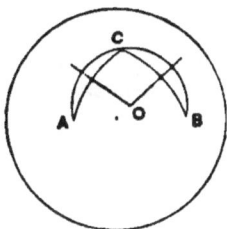

PROBLEM II

To pass a great circle through any two points on the surface of a sphere.

Let A and B be any two points on the surface of a sphere. Make either point, as A, a pole, and from it describe a circumference of either a small or a great circle; and from the other point, as B, pass the arc of a great circle cutting the circumference CD at right angles; it will pass through the point A. The pole of the great circle must be taken at a quadrant's distance from B and from E.

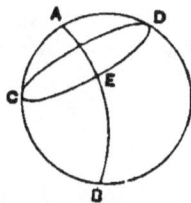

REM.—The point on the circumference of the circle through which the arc of a great circle passes, is found by Prob. 2, Book I, and from the pole of this point and B, a quadrant's distance from each, the arc is drawn.

THEOREM IX.

In every spherical triangle, the sum of any two sides is greater than the third side.

Let ABC be a spherical triangle; then will AC + BC > AB. Join the vertices A, B, and C, with O, the center of the sphere, and a triedral angle is formed, the arcs of whose facial angles are the sides of the spherical triangle; but (Th. 1) "the sum of any two of the plane angles which form a triedral

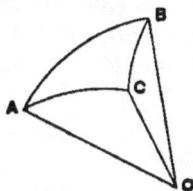

angle is greater than a third; hence, the sum of any two sides of a spherical triangle is greater than the third side.

THEOREM X.

The sum of the three sides of a spherical triangle is less than the circumference of a great circle.

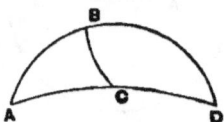

Let ABC be a spherical triangle. Produce AB and AC until they meet in D. The arcs ABD and ACD are semi-circumferences, since two great circles always bisect each other.

In the triangle BCD, the sum of the two sides CD and BD is greater than the third side BC; hence the sum of the three sides AB, AC, and BC is less than the circumference of a great circle.

REM.—The sides of a spherical triangle are arcs which correspond with and measure the facial angles of a triedral angle, the vertex of which is at the center of the sphere.

THEOREM XI.

The sum of all the sides of a spherical polygon is less than the circumference of a great circle.

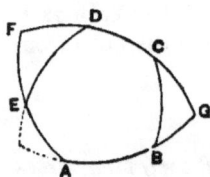

Let ABCDE be a spherical polygon. Produce AB and DC until they meet in G; also produce AE and CD until they meet in F. BC is less than BG + CG, and DE is less than EF + FD; hence the sum of the sides of the polygon is less than the sum of the sides of the triangle AFG; and the sum of the sides of the triangle is less than the circumference of a great circle (Th. 10); much more then is the sum of all the sides of the polygon less than the circumference of a great circle.

THEOREM XII.

If from the vertices of the angles of a spherical tri-angle as poles, with a distance between the points of the dividers equal to a quadrant, arcs be drawn forming another spherical triangle, the vertices of this triangle will be respectively the poles of the sides of the first triangle.

Let ABC be a spherical triangle, and then with each vertex as a center and a distance between the points of the dividers equal to ninety degrees, describe the polar triangle DEF. First with A as a pole describe the arc EF, with B as a pole describe DF, and with C as a pole describe DE; in each case the distance between the points of the dividers being 90 degrees. Since AE = 90° and CE = 90°, E is the pole of AC ; and since BD = 90° and DC = 90°, D is the pole of BC ; and as BF = 90° and AF = 90°, F is the pole of AB.

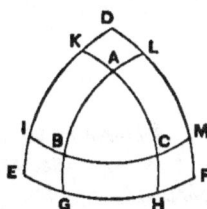

THEOREM XIII.

Any angle in one of two polar triangles is measured by a semi-circumference minus the side opposite of the other triangle.

Let ABC and DEF be triangles polar to each other ; produce the sides of ABC until they meet those of DEF; A is the pole of the arc GH by which the angle A is meas-ured, E is the pole of KH, and F is the pole of LG ; hence, EH = 90° and FG = 90° ; hence, GH = 180°—EF ; that is, the angle A is measured by a semi-circumference minus the side opposite in its polar triangle ; so also in regard to each of the other angles.

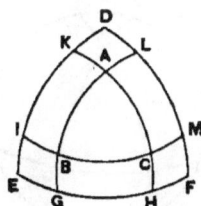

Cor. to 10th and 13th Th.—The sum of the three angles of a spherical triangle is less than six right angles and greater than two.

Cor. 2.—A spherical triangle may have two or even three right angles, or as many obtuse angles; when it has three right angles it is called the trirectangular triangle, whose surface is equal to one-eighth the surface of the sphere.

Cor. 3.—The sum of the three angles of a spherical triangle is not a constant quantity, but varies between two and six right angles, never reaching either of these limits.

Rem.—The excess of the sum of the angles of a spherical triangle over two right angles is called the spherical excess.

THEOREM XIV.

A lune is to the surface of a sphere as the arc which measures its angle is to the circumference of a great circle.

Surface of lune : 8 trirectangular tri. :: angle A : 4;

$$\therefore \quad 4L = A \times 8T, \quad \text{and} \quad L = A \times 2T.$$

Cor.—The surface of a lune is equal to its angle multiplied by twice the trirectangular triangle.

Rem.—As the right angle is the unit, the angle will be indicated by a fraction; as $\frac{4\cdot 3}{9\cdot 0}$, which is 43 degrees.

THEOREM XV.

The area of a spherical triangle is equal to its spherical excess multiplied by the trirectangular triangle.

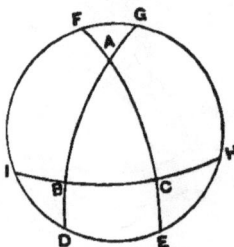

Let ABC be the spherical triangle.

The triangles ADE and AFG form a lune
     = angle A × 2T;
the triangles BGH and BID form a lune
     = angle B × 2T;
the triangles CFI and CEH form a lune
     = angle C × 2T.

By addition we get

$$2 (A + B + C) T = 2 \text{ area of ABC} + 4T,$$

and
$$(A + B + C) T = \text{area ABC} + 2T,$$

and
$$(A + B + C) T - 2T = \text{area ABC}.$$

$$\text{Area ABC} = (A + B + C - 2) T.$$

THEOREM XVI.

The area of a spherical polygon is equal to its spherical excess multiplied by the trirectangular triangle.

Joining A and C and A and D by arcs of great circles, we divide the polygon into triangles.

The sum of all the angles of the triangles is equal to the sum of all the angles of the polygon ; hence, the area of the polygon is equal to the sum of the areas of the triangles.

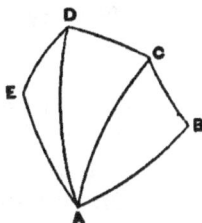

Let $S =$ Sum of all the angles,

and $n =$ Number of the sides of the polygon,

$(n - 2) =$ "   " triangles.

Area of polygon $= [S - (n - 2) \, 2] \, T$;

Area of ABCDE $= (S - 2n + 4) \, T$.

APPLICATION OF LOGARITHMS.

A **Logarithm** is the exponent or power of some number, which is called the base of the logarithms.

The base of the logarithms in common use is 10; hence,

$10^0 = \frac{10}{10} = 1$; therefore, the logarithm of 1   is 0.

$10^1 = 10$;   "   "   "   " 10   " 1.

$10^2 = 100$;   "   "   "   " 100   " 2.

$10^3 = 1000$;   "   "   "   " 1000 " 3.

The integral number of the logarithm is called the **Characteristic,** and when positive is one less than the number of integral figures in the number of which it is the logarithm.

The logarithm of any number between 1 and 10 must be between 0 and 1; the logarithm of any number between 10 and 100 must be between 1 and 2; that is, the logarithm of any number between 1 and 10 is a fraction, and the logarithm of any number between 10 and 100 is 1 + a fraction.

The calculations of the fractions is made by an algebraic process, and logarithmic tables are formed to facilitate trigonometrical computations. Observe also that

$$.1 \quad = \frac{1}{10} = 10^{-1}; \text{ that is, the logarithm of .1 is } -1;$$

$$\text{and} \quad .01 \ = \frac{1}{10^2} = 10^{-2}; \text{ " " " " " .01 " } -2.$$

$$.001 = \frac{1}{10^3} = 10^{-3}; \text{ " " " " " .001 " } -3.$$

REM.—The characteristic of the logarithm of a fraction is negative and corresponds to the number of decimals; hence, the logarithm of any number between 1 and .1 is between 0 and -1, and is put -1 + a fraction.

As logarithms are exponents, they can only be used as such; that is, for Multiplication and Division, for Involution and Evolution; thus, $a^1 \times a^1 = a^2$; $a^2 \times a^3 = a^5$; $a^1 \times b^2 = ab^2$; $a^2 \times b^2 = a^2b^2$; $a^1 \times a^1 \times a^1 = a^3$; $a^3 \div a^2 = a^1$; $\sqrt{a^2} \times \sqrt{a} = \sqrt{a^3} = a^{\frac{3}{2}}$; $\sqrt{a^2b} = ab^{\frac{1}{2}}$; $\sqrt{a^2b^3c^5} = ab^{\frac{3}{2}}c^{\frac{5}{2}}$; $a^2b^3c^4d^5 \div ab^4c^3d^6e^2 = ab^{-1}cd^{-1}e^{-2}$.

REM.—Multiplication is performed by adding the exponents of the multiplier and multiplicand; Division by subtracting the exponents of the divisor from those of the dividend; Involution by multiplying the exponent by the exponent of the power to which it is to be raised; and Evolution by dividing the exponents by the index of the root.

From the table of logarithms the decimal parts are found, and the characteristic must be added. The characteristic of an integral number will always be one less than the number of integral figures.

The logarithm of	4532	is	3.656290
" " "	453.2	"	2.656290
" " "	45.32	"	1.656290
" " "	4.532	"	0.656290
" " "	.4532	"	$\bar{1}$.656290
" " "	.04532	"	$\bar{2}$.656290

To find the logarithm of a number of three or less figures, find the given number in the first column of the table, and take the logarithm under zero; if there be four figures, reserve the one in the units' place and find the other three in the first column, and take the logarithm under the figure corresponding to the reserved figure.

The logarithm of	4530	is	3.656098
" " "	45300	"	4.656098
" " "	453000	"	5.656098

To find the logarithm of a number which has more than four figures:

First find the logarithm of the number expressed by the four left-hand figures; then multiply the common difference given in the table by all the figures left, and of this product reject as many figures on the right as there were reserved figures, and add the balance to the logarithm already found.

Thus, find the logarithm of 564236.

The characteristic is 5.

The decimal part of the logarithm of 5642 is   .751433
Common difference, $77 \times 36 = 27$) 72   28
                             .751461

As the figures cut off are more than .5, we add 28 instead of 27. To which add the characteristic and the

logarithm of 564236 = 5.751461.

To find the number corresponding to the given logarithm :

If the given logarithm be in the table, take the three figures opposite in the first column and the one immediately over it at the top of the page, and point off one more integral figure than there are in the characteristic of the logarithm.

But if the given logarithm is not in the table, take the number corresponding to the next less logarithm ; then divide the difference between this logarithm and the given one by the tabular difference, annexing ciphers to the dividend, and then affixing this quotient to the number already found. The figure occupying the tenths' place must take the first place after the number already taken from the table; this will sometimes be 0.

EXAMPLES.

1. Find the product of 4573 and 6321.

logarithm 4573 =	3.660201
logarithm 6321 =	3.800786
Next less number (2890),	7.460987
Next less logarithm,	898
Difference,	89
Com. diff., 150.	150) 890 (5933

Hence, number corresponding, 28905933.

2. Divide 1728 by 12.

logarithm 1728 =	3.237544
logarithm 12 =	1.079181
	2.158363

Number corresponding,   144.

There will be slight inaccuracies, as the logarithms are only carried to six places of decimals.

3. Find the logarithmic sine of 37° 15′ 25″.

sin 37° 15′ =	9.781966
sin 25″ (diff. 2.77 × 25 = 69.25) =	69
sin 37° 15′ 25″ =	9.782035

4. Find the tangent of 57° 30′ 30″.

tang 57° 30′ =	10.195813
tang 30″ (4.65 × 30 = 139.50) =	139
tang 57° 30′ 30″ =	10.195952

I will not add any more examples, as the learner will soon become familiar with the use of the tables.

TRIGONOMETRY.

Trigonometry treats of the methods of finding the unknown parts of a triangle, when certain parts are known.

Every triangle has six parts, viz., three sides and three angles. When three of these are known, one at least being a side, the other three can be found by these methods.

Plane Trigonometry treats of plane triangles.

When one thing depends upon another for its value, the first is said to be a **function** of the second.

Sines, tangents, cosines, cotangents, etc., are functions of arcs or angles.

DEFINITIONS.

1. The **Sine** of an angle or arc is a line drawn from the end of the arc perpendicular to the radius drawn from the beginning of the arc to its center, or to the vertex of the angle.

2. The **Versed Sine** is the distance from the foot of the sine to the beginning of the arc.

3. The **Cosine** is the sine of the complement of the angle ; it is equal to the radius minus the versed sine.

4. The **Tangent** is a line drawn from the beginning of the arc perpendicular to the radius at its extremity, and is limited by a line drawn from the vertex of the angle through the point at the end of the arc.

5. The line joining the vertex of the angle and the extremity of the tangent is called the **Secant.**

6. The **Cotangent** is the tangent of the complement of the angle.

The functions of an angle are exhibited in the following diagram:

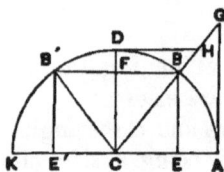

BE is the sine of the acute angle ACB, and the cosine of BCD, its complement.

BF or CE is the cosine of the acute angle ACB, and the sine of BCD, its complement.

AG is the tangent of the acute angle ACB, and the cotangent of BCD, its complement.

DH is the cotangent of the acute angle ACB, and the tangent of BCD, its complement.

The radius DC is the sine of the right angle ACD; its cosine is zero; the tangent would be AG extended until it meet CD extended; but they are parallel; hence the tangent of 90° is infinite, and the cotangent zero.

If the angle is obtuse, as ACB', its sine is B'E', the same as of B'CK, its supplement. All the other functions, except the versed sine of the obtuse angle, will be the same length as of the acute angle, its supplement.

All the functions of acute or right angles or arcs terminating in the first quadrant have plus signs. If the angle is obtuse, its cosine, tangent, and cotangent will have a minus sign.

Cor. 1.—The sine of an angle cannot be greater than the radius; but the tangent may have any value whatever.

Cor. 2.—It is needless to say that the cosine becomes equal to the radius, or the cotangent infinite, as this only happens when there is no angle.

PROBLEM I.

To show the relations of the sines, cosines, tangents, and cotangents of the angles of a right-angled triangle with the sides of the triangle,

Let PHB be a triangle, right-angled at H; PF or PC is the radius of the arc CF; DC is the sine of the angle BPH, PD its cosine, and FE its tangent and cotangent of HBP, its complement.

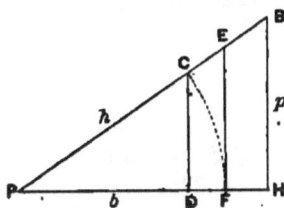

The triangles PDC and PHB are similar; hence the proportions,

CD : BH :: PC : PB;   and   PD : PH :: PC : PB;
that is,   that is,

$$\text{sine } P : p :: R : h ; \quad (1) \qquad \text{sine } B : b :: R : h ; \quad (2)$$

$$\therefore p = \frac{\text{sine } P \times h}{R}, \ (1) \qquad \therefore b = \frac{\text{sine } B \times h}{R}, \ (1)$$

$$\text{and} \quad h = \frac{p \times R}{\sin P}, \quad (2) \qquad \text{and} \quad h = \frac{b \times R}{\sin B}, \quad (2)$$

$$\text{and} \quad \text{sine } P = \frac{p \times R}{h} \quad (3) \qquad \text{sine } B = \frac{b \times R}{h}$$

$$= \text{cosine } B, \quad (3) \qquad\qquad = \text{cosine } P, \quad (3)$$

and from 2 and 2,

$$\frac{p \times R}{\sin P} = \frac{b \times R}{\sin B}, \quad \text{and} \quad \frac{p}{\sin P} = \frac{b}{\sin B};$$

$$\therefore \ \sin P : \sin B :: p : b. \quad (4)$$

Alternating 1st and 2d proportions,

$$\sin P : R :: p : h, \quad \text{and} \quad \sin B : R :: b : h. \quad (4)$$

The triangles PFE and PHB are similar; hence the proportion,

$$EF : BH :: PF : PH; \text{ that is, tang } P : p :: R : b;$$

hence, $p = \dfrac{\text{tang } P \times b}{R}$,   and $b = \dfrac{\text{tang } B \times p}{R}$, by analogy. (5)

Tang $P = \dfrac{p \times R}{b} = \cot B$, and tang $B = \dfrac{b \times R}{p} = \cot P$, an., (6)

and $p = \dfrac{R \times b}{\text{tang } B}$,   and $b = \dfrac{R \times p}{\text{tang } P}$.   (5)

REM.—In a right-angled triangle, either acute angle is a complement of the other ; hence, the sine of the one is the cosine of the other, and the tangent of the one is the cotangent of the other.

As these formulas are general principles of right-angled triangles ; they must be fixed indelibly on the mind of the student.

REM.—The small letters represent the sides opposite the angles having corresponding large letters.

FORMULAS.

· 1. Either side is equal to the sine of the opposite angle multiplied by the hypothenuse and divided by the radius.

2. The hypothenuse is equal to the radius multiplied by either side, and divided by the sine of the angle opposite this side.

3. The sine of either acute angle is equal to the opposite side multiplied by the radius and divided by the hypothenuse.

4. Either side, including the hypothenuse, is to any other side as the sine of the angle opposite the former side is to the sine of the angle opposite the latter side.

5. Either side is equal to the tangent of its opposite angle multiplied by the other side and divided by the radius ; or,

Either side is equal to the radius multiplied by the other side and divided by the tangent of the angle opposite this other side.

6. The tangent of either acute angle is equal to its opposite side, multiplied by the radius and divided by the other side.

EXAMPLES.

1. Given $h = 43$, $B = 25°$, to find p, b, and P.
$$P = 90° - 25° = 65°.$$

Formula 1, $p = \dfrac{\sin P \times h}{R}$; $\therefore \log p = \log \sin 65° + \log 43 - \log R$,

log sin of 65°	= 9.957276
log 43 — 10	= 1.633468
$p = 38.971$, No. corres.	1.590744

Formula 1, $b = \dfrac{\sin B \times h}{R}$; $\therefore \log. b = \log \sin 25° + \log 43 - \log R$,

log sin 25°	= 9.625948
log 43 — 10	= 1.633468
$b = 18.172$, No. corres. to log,	1.259416

2. Given $h = 624$, $P = 48°$, to find p, b, and B.
$$B = 42°, \ p = 463.723, \text{ and } b = 417.538, \ Ans.$$

3. Given $b = 535$, B $= 65° 15'$, to find P, p, and h.

      P $= 24° 45'$, $p = 246.738$, and $h = 589.114$, *Ans.*

4. Given $b = 47$, P $= 35°$, to find B, p, and h.

      ·B $= 55°$, $p = 32.91$, and $h = 57.376$, *Ans.*

5. Given $p = 275$, B $= 58°$, to find P, b, and h.

      P $= 32°$, $b = 440.92$, and $h = 518.946$, *Ans.*

6. Given $p = 15$, $b = 25$, to find P, B, and h. (Use formula 6.) $h = 29.05$, B $= 59° 23' 3''$, and P $= 30° 36' 57''$, *Ans.*

PROBLEM II.

To show that the sines of the angles, in any plane triangle, are respectively proportional to their opposite sides.

Let ABC be any plane triangle, and from any vertex as B draw a perpendicular BD to the opposite side AC, making two right-angled triangles, ABD and BCD; from which we have in the triangle ABD, BD $= \dfrac{\sin A \times c}{R}$, and in the

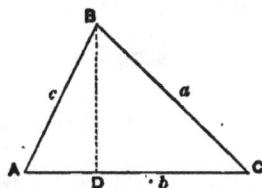

triangle BCD, BD $= \dfrac{\sin C \times a}{R}$; $\therefore \sin A \times c = \sin C \times a$; hence, $\sin A : \sin C :: a : c$, and by inversion, $\sin C : \sin A :: c : q$, and $\sin B : \sin A :: b : a$, and $\sin A : \sin B :: a : b$; hence, in any triangle, the sines of the angles are respectively proportional to their opposite sides.

EXAMPLES.

1. Given A $= 55°$, B $= 51°$, $c = 143$, to find C, a, and b.

      C $= 74°$, $a = 121.86$, and $b = 151.61$, *Ans.*

2. Given A $= 60$, $c = 54$, and A $= 26°$, to find B, C, and b.

C $= 23° 10' 13''$, B $= 130° 45' 47''$, and $b = 103.667$, *Ans.*

REM.—If c is greater than a there is two triangles; thus, $c = 60$, $a = 54$, and A $= 26°$.

Observe, that the angle AC'B is the supplement of ACB.

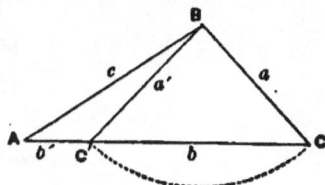

·C $= 29° 8' 56''$, C' $= 150° 51' 4''$, B $= 124° 51'.4''$, B' $= 3° 8' 56''$, $b = 101.089$, and $b' = 6.7665$, *Ans.*

PROBLEM III.

To show by Diagram No. 2 certain relations of the functions of angles, when the radius is unity.

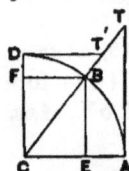

In the triangle CEB,

$$\overline{BE}^2 + \overline{CE}^2 = \overline{CB}^2;$$

that is, $\quad \sin^2 C + \cos^2 C = R^2 = 1, \qquad (1)$

$$\sin^2 C = 1 - \cos^2 C, \qquad (2)$$

and $\qquad \cos^2 C = 1 - \sin^2 C. \qquad (3)$

The triangles CEB and CAT are similar, and also CFB and CDT′ are similar.

∴ CE : CA :: EB : AT, $\qquad$ ∴ CF : CD :: FB : DT′,

cos C : 1 :: sin C : tang C, $\qquad$ sin C : 1 :: cos C : cot C,

cos C × tang C = sin C, (4) $\qquad$ sin C × cot C = cos C, (6)

$$\text{tang } C = \frac{\sin C}{\cos C}; \quad (5) \qquad \cot C = \frac{\cos C}{\sin C}, \quad (7)$$

$$\text{tang } C \times \cot C = \frac{\sin C}{\cos C} \times \frac{\cos C}{\sin C} = 1;$$

$$\therefore \quad \tan C = \frac{1}{\cot C}, \qquad (8)$$

$$\cot C = \frac{1}{\text{tang } C}. \qquad (9)$$

REM.—In the above diagram, C represents any angle; hence the relations apply to all angles.

SYNOPSIS OF THE FORMULAS.

$$\sin^2 + \cos^2 = 1. \qquad (1)$$

$$\sin^2 = 1 - \cos^2. \qquad (2)$$

$$\cos^2 = 1 - \sin^2. \qquad (3)$$

$$\sin = \cos \times \text{tang.} \qquad (4)$$

$$\text{tang} = \frac{\sin}{\cos}. \qquad (5)$$

$$\cos = \sin \times \cot. \qquad (6)$$

$$\cot = \frac{\cos}{\sin}. \qquad (7)$$

$$\text{tang} = \frac{1}{\cot}. \qquad (8)$$

$$\cot = \frac{1}{\text{tang}}. \qquad (9)$$

PROBLEM IV.

To show the relations of the functions of a right-angled triangle with the sides, when the radius is unity.

The triangles PDC and PHB are similar, and give the proportions,

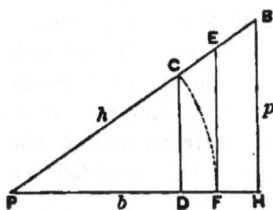

CD : BH :: PC : PB,

and   PD : PH :: PC : PB.

$\sin P : p :: 1 : h,$

$\sin B : b :: 1 : h.$

$p = \sin P \times h,$ $b = \sin B \times h,$ (1)

$h = \dfrac{p}{\sin P},$ $h = \dfrac{b}{\sin B},$ (2)

$\sin P = \dfrac{p}{h},$ $\sin B = \dfrac{b}{h}.$ (3)

From (2), $\dfrac{p}{\sin P} = \dfrac{b}{\sin B},$ $\therefore \ \sin P : \sin B :: p : b.$

The triangles PFE and PHB are similar;

$\therefore$ BH : EF :: PH : PF,

$p : \text{tang } P :: b : 1;$

hence, $p = \text{tang } P \times b,$

$b = \dfrac{p}{\text{tang } P};$

and by a similar process,   (4)

$b = \text{tang } B \times p,$

$p = \dfrac{b}{\text{tang } B};$

$\text{tang } P = \dfrac{p}{b} = \cot B,$ (5)

and $\text{tang } B = \dfrac{b}{p} = \cot P.$

ENUNCIATION OF FORMULAS.

1. Either side is equal to the product of the sine of the opposite angle and the hypothenuse.

2. The hypothenuse is equal to either side, divided by the sine of the angle opposite that side.

3. The sine of either acute angle is equal to its opposite side divided by the hypothenuse.

4. Either side is equal to the product of the tangent of its opposite angle and the other side; or, either side is equal to the other side divided by the tangent of the angle opposite that other side.

5. The tangent of either angle is equal to its opposite side divided by the other side.

REM.—In computing these formulas by logarithms, when the formula consists of the product of two numbers, 10 must be subtracted from the sum of the logarithms; and when it is fractional, 10 must be added to the difference of the logarithms.

PROBLEM V.

To find the sine and cosine of the sum and difference of two arcs, whose sines and cosines are known.

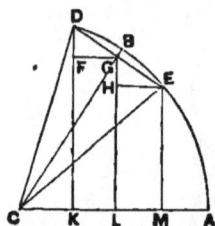

Let ACB and BCD be the two angles whose sines and cosines are known.

Let the angle ACB be designated angle A.
     "   "   BCD "   "   "   B.

Make   BCE = BCD;

then   DK = $\sin (A + B)$,

and   EM = $\sin (A - B)$.

CK = $\cos (A + B)$,

and   CM = $\cos (A - B)$.

The triangles DFG and GHE are equal and similar to CLG, whose sides are respectively perpendicular to those of DFG and GHE.

∴   the angle FDG = angle LCG = angle A.

In the triangle DFG.

$$DF = DG \times \cos A,$$

and in the triangle CLG,

$$GL = CG \times \sin A.$$
$$DF = \sin B \times \cos A,$$

and

$$GL = \cos B \times \sin A,$$

$$GL + DF = DK = \sin (A+B) = \sin A \times \cos B + \cos A \times \sin B$$
$$GL - DF = EM = \sin (A-B) = \sin A \times \cos B - \cos A \times \sin B$$

By addition,

$$\sin (A + B) + \sin (A - B) = 2 \sin A \times \cos B,$$

and

$$\sin (A + B) - \sin (A - B) = 2 \cos A \times \sin B.$$

Put $A + B = M$, and $A - B = N$;

then $A = \tfrac{1}{2}(M + N)$, and $B = \tfrac{1}{2}(M - N)$;

then

$$\sin M + \sin N = 2 \sin \tfrac{1}{2}(M+N) \times \cos \tfrac{1}{2}(M-N), \quad (1)$$

and

$$\sin M - \sin N = 2 \cos \tfrac{1}{2}(M+N) \times \sin \tfrac{1}{2}(M-N); \quad (2)$$

dividing (1) by (2), and reducing by $\tan = \dfrac{\sin}{\cos}$,

$$\frac{\sin M + \sin N}{\sin M - \sin N} = \frac{\sin \tfrac{1}{2}(M+N) \times \cos \tfrac{1}{2}(M-N)}{\cos \tfrac{1}{2}(M+N) \times \sin \tfrac{1}{2}(M-N)} = \frac{\tan \tfrac{1}{2}(M+N)}{\tan \tfrac{1}{2}(M-N)} \quad (3)$$

In the triangle CLG, $CL = CG \times \cos A = \cos B \times \cos A$.

" " DFG, $FG = DG \times \sin A = \sin B \times \sin A$.

By subtraction and addition,

$$CL - FG = CK = \cos M = \cos A \times \cos B - \sin A \times \sin B$$
$$CL + FG = CM = \cos N = \cos A \times \cos B + \sin A \times \sin B$$

Cor.—By addition and subtraction,

$$\cos M + \cos N = 2 \cos \tfrac{1}{2}(M + N) \times \cos \tfrac{1}{2}(M - N), \quad (4)$$

$$\cos M - \cos N = -2 \sin \tfrac{1}{2}(M + N) \times \sin \tfrac{1}{2}(M - N); (5)$$

dividing (5) by (4),

$$\frac{\cos M - \cos N}{\cos M + \cos N} = \frac{-\sin \tfrac{1}{2}(M + N)}{\cos \tfrac{1}{2}(M + N)} \times \frac{\sin \tfrac{1}{2}(M - N)}{\cos \tfrac{1}{2}(M - N)}$$
$$= -\tan \tfrac{1}{2}(M + N) \times \tan \tfrac{1}{2}(M - N).$$

Observe that the tangent has a negative sine, which is correct, as the numerator of the first member of the equation is negative, the cosine N being greater than the cos M; a small angle has a larger cosine than a large angle.

EXAMPLES.

1. To find the sine and cosine of 30°, 60°, and 45°, when the radius is 1.

The sine of 30° is half the chord of 60°, the chord of 60° is radius $= 1$; hence, sine of 30° $= \frac{1}{2}$, and cosine of 60° $= \frac{1}{2}$.

$$\therefore \quad \sin 30° = \cos 60° = \tfrac{1}{2}.$$

Cosine of 30° $= \sqrt{1 - \tfrac{1}{4}} = \sqrt{\tfrac{3}{4}} = \tfrac{1}{2}\sqrt{3}$; sine of 60° $= \tfrac{1}{2}\sqrt{3}$;

and $\qquad\qquad \cos 30° = \sin 60° = \tfrac{1}{2}\sqrt{3}.$

When the angle is 45°, the sine and cosine will be equal, and as $\sin^2 + \cos^2 = 1$, that is, $\tfrac{1}{2} + \tfrac{1}{2} = 1$.

$$\sin^2 45° = \tfrac{1}{2}, \quad \text{and} \quad \sin 45° = \sqrt{\tfrac{1}{2}} = \tfrac{1}{2}\sqrt{2} = \cos 45°,$$

and $\qquad\qquad \sin 45° = \cos 45° = \tfrac{1}{2}\sqrt{2}.$

2. To find the sine, cosine, tangent, and cotangent of every arc from 1° to 90°.

The semi-circumference, when the radius is 1, is 3.1415926535; which being divided by 10800, the number of minutes it contains, gives the length of 1′, equal to .0002908882, which in so small an arc does not differ materially from the sine of 1′, and may be regarded as such; and $\cos 1' = \sqrt{1 - \sin^2 1'} = .9999999577.$

By taking the formula from Prob. 5,

$$\sin (A + B) + \sin (A - B) = 2 \sin A \times \cos B,$$

by transposition,

$$\sin (A + B) = 2 \sin A \times \cos B - \sin (A - B),$$

and making $B = 1'$, and $A = 1'$, $2'$, $3'$, $4'$, etc., in succession,

$$\sin 1' = \qquad\qquad\qquad\qquad .0002908882;$$
$$\sin 2' = 2 \sin 1' \times \cos 1' - \sin 0 = .0005817764;$$
$$\sin 3' = 2 \sin 2' \times \cos 1' - \sin 1' = .0008726646.$$

By continuing this process, we can get the sines of every arc from 1' to 90°; and taking them in an inverse order, we have the cosines of every arc from 1' to 90°; then, as $\text{tang} = \dfrac{\sin}{\cos}$, we can get the tangents of every arc from 1' to 90°; and by taking them in an inverse order we have the cotangents of every arc from 1' to 90°. These will form a table of natural sines, cosines, tangents and cotangents.

The logarithms of these numbers, with the addition of 10 to to each logarithm, forms the table of logarithmic sines, cosines, etc. In the table the radius is taken as ten billions, whose logarithm is 10; and as the functions are proportional to the radii, hence the natural sines, etc., must be multiplied by this number, which is done by adding the logarithms of the natural sine, etc., and of the radius.

PROBLEM VI.

Two sides and the included angle of a triangle given, to find the other angles.

By Problem 2,

$$a : b :: \sin A : \sin B;$$

by composition and division,

$$a + b : a - b :: \sin A + \sin B : \sin A - \sin B,$$

$$\therefore \quad \frac{a + b}{a - b} = \frac{\sin A + \sin B}{\sin A - \sin B}.$$

By Problem 5, Formula (3),

$$\frac{\sin M + \sin N}{\sin M - \sin N} = \frac{\text{tang} \frac{1}{2} (M + N)}{\text{tang} \frac{1}{2} (M - N)};$$

$$\therefore \quad \frac{a + b}{a - b} = \frac{\sin A + \sin B}{\sin A - \sin B} = \frac{\text{tang} \frac{1}{2} (A + B)}{\text{tang} \frac{1}{2} (A - B)};$$

hence the proportion,

$$a + b : a - b :: \text{tang} \tfrac{1}{2} (A + B) : \text{tang} \tfrac{1}{2} (A - B).$$

Knowing the sum and difference of two angles, we easily find the angles.

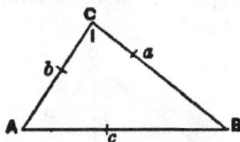

5

PROBLEM VII.

To find the area of a triangle, having given two sides and the included angle.

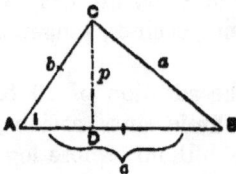

The angle A and the sides b and c given.

$$\text{area } ABC = \tfrac{1}{2}pc.$$

In the triangle ADC,

$$p = \sin A \times b;$$

$\therefore$　　area of ABC $= \tfrac{1}{2}b \times c \times \sin A.$

PROBLEM VIII.

If from the vertex of any angle of a triangle a line be drawn perpendicular to the opposite side, produced if necessary, then will the sum of the segments of the opposite side be to the sum of the other two sides as the difference of those sides is to the difference of the segments.

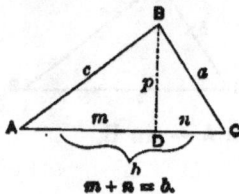

$$p^2 = a^2 - n^2 = c^2 - m^2,$$
$$m^2 - n^2 = c^2 - a^2,$$
$$(m + n)(m - n) = (c + a)(c - a),$$
$$m + n : c + a :: c - a : m - n.$$

PROBLEM IX.

If from the half sum of the three sides of a triangle, each side be subtracted separately, then the square root of the continued product of the half sum and the three remainders will be the area of the triangle.

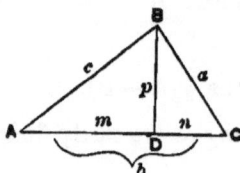

$$p^2 = c^2 - m^2 = a^2 - n^2,$$

and

$$c^2 - a^2 = m^2 - n^2,$$
$$m^2 - n^2 = c^2 - a^2,$$
$$(m + n)(m - n) = c^2 - a^2;$$

and as
$$m + n = b,$$
$$b\,(m - n) = c^2 - a^2,$$
$$m - n = \frac{c^2 - a^2}{b},$$
$$m + n = b,$$

$$\overline{\qquad\qquad}$$

$$2m = b + \frac{c^2 - a^2}{b},$$

$$m = \frac{b^2 + c^2 + a^2}{2b},$$

and
$$p^2 = c^2 - m^2 = c^2 - \left(\frac{b^2 + c^2 - a^2}{2b}\right)^2,$$

$$p = \sqrt{\frac{4b^2c^2 - (b^2 + c^2 - a^2)^2}{4b^2}}$$

$$= \sqrt{\frac{(2bc + b^2 + c^2 - a^2)\,[2bc - (b^2 + c^2 - a^2)]}{4b^2}}$$

$$= \sqrt{\frac{[(b + c)^2 - a^2] \times [a^2 - (b - c)^2]}{4b^2}}$$

$$= \sqrt{\frac{(b + c + a)\,(b + c - a)\,(a + b - c)\,(a + c - b)}{4b^2}}$$

area of ABC $= \frac{1}{2}pb$
$$= \sqrt{\frac{(b + c + a)\,(b + c - a)\,(a + b - c)\,(a + c - b)}{4b^2}} \times \sqrt{\frac{b^2}{4}};$$

b^2 is canceled, and the two 4's factored,

area of ABC $= \sqrt{\dfrac{(b + c + a)\,(b + c - a)\,(a + b - c)\,(a + c - b)}{2 \quad\quad 2 \quad\quad 2 \quad\quad 2}}.$

EXAMPLES.

Two sides and the included angle given.

1. Given $a = 75$, $b = 90$, and $C = 20°$, to find A, B, and e.

$b + a : b - a :: \text{tang. } \frac{1}{2}\,(B + A) : \text{tang. } \frac{1}{2}\,(B - A);$
$\qquad 165 : 15 :: \text{tang. } 80° : \text{tang. } B - A.$

$\qquad \text{tang. } \frac{1}{2}\,(B - A) = \log 15 + \log \text{tang. } 80° - \log 165,$
$\qquad\qquad \frac{1}{2}\,(B + A) = \quad 80° \ 00' \ 00'',$
$\qquad\qquad \frac{1}{2}\,(B - A) = \quad 27 \ 16 \ 27.$

$$\overline{\qquad\qquad\qquad\qquad}$$

$\qquad\qquad\qquad B = 107° \ 16' \ 27'',$
$\qquad\qquad\qquad A = \quad 52° \ 43' \ 33''.$

$$\text{Sin A} : \sin C :: a : c,$$

$$\log c = \log \sin C + \log a - \log \sin A.$$

Arith. comp. 165	=	7.782516
log 15	=	1.176091
log tang. 80°	=	10.753681
tang. $\frac{1}{2}$ (B—A) = 27° 16′ 27″ =		9.712288

Arith. comp. sin 52° 43′ 33″	=	0.099225
log sin C 20°		9.534052
log a 75		1.875061
$c = 32.235$		1.508338

When the three sides are given.

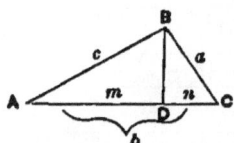

2. Given $a = 237$, $b = 495$,   and $c = 327$.

$$m + n : c + a :: c - a : m - n;$$
$$m + n = b,$$
$$\log (m - n) = \log (c + a) + \log (c - a)$$
$$- \log b.$$

:	$m + n$	= 495
	$m - n$	= 102.546
	m	= 298.773,
	and n	= 196.227.

In the triangle ABD,

$$c : m :: R : \sin ABD ;$$

$$\log \sin ABD = \log R + \log m - \log c.$$

Arith. comp. 495	=	7.305395
log 564	=	2.751279
log 90	=	1.954243
$(m - n) = 102.546$		2.010917

Arith. comp. 327		7.485452
log 298.773		2.475342
log R		10.000000
sin ABD = cos A 23° 59′		9.960794

In the triangle BDC, $a : n :: R : \cos C$;

$$\log \cos C = \log n + \log R - \log a.$$

$$A = 23° \ 59' \ 00''$$
$$C = 34° \ \ 6' \ 36''$$
$$180 - \overline{58° \ \ 5' \ 36''} = 121° \ 54' \ 24'' = B.$$

Arith. comp. 237 =	7.625252
log 196.227 =	2.292759
log R	10.000000
cos C 34° 6' 36''	9.918011

PROBLEM.

A side and two adjacent angles given, also two sides and an angle opposite one of them.

EXAMPLES.

1. Given A = 32°, a = 40, and b = 50, to find B, C, and c.

Ans. $\begin{cases} B = \ \ 41° \ 28' \ 59'', \ C = 106° \ 31' \ \ 1'', \ \text{and} \ c = 72.368. \\ B = 138° \ 31' \ \ 1'', \ C = \ \ \ 9° \ 28' \ 59'', \ \text{``} \ \ c = 12.436. \end{cases}$

In this case there are two triangles.

2. Given a = 450, b = 540, and C = 80°, to find A, B, and c.
A = 43° 49', B = 56° 11', and c = 640.08.

3. Given a = 40, b = 34, and c = 25 yards, to find the angles. A = 83° 53' 16'', B = 57° 41' 24'', and C = 38° 25' 20''.

4. Given b = 306, c = 274, and B = 78° 13', to find A, C, and a. A = 40° 33', C = 61° 14', and a = 203.2.

5. Given B = 100°, a = 280.3, and c = 304, to find A, C, and b. A = 38° 3' 3'', C = 41° 56' 57'', and b = 447.856.

6. Find the area of a triangle having two sides equal to 30 and 40 ft. respectively, and the included angle 28° 57',

Ans. 290.427 sq. ft.

7. Find the area of a triangle whose sides are respectively 30, 40, and 50 rods. *Ans.* 3 acres 3 rods.

8. What is the area of a triangle, whose base is 50 rods and altitude 30 rods? *Ans.* 4 acres 2 rods 30 perches.

PRACTICAL PROBLEMS.

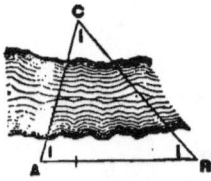

1. Find the distance AC across a deep river, having given AB = 500 yards, the angle BAC = 74° 14′, and the angle ABC = 49° 23′.

$$\sin C : \sin B :: c : b;$$
$$\log b = \log \sin B + \log c - \log \sin C$$
$$= 577.8 \text{ yards.}$$

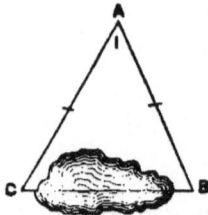

2. Given AC = 735 yards, BC = 840, and the angle C = 55° 40′, to find AB.

Two sides and the included angle.

$$AB = 741.$$

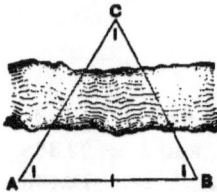

3. Given AB = 600 yards, and the adjacent angles A = 57° 35′ and B = 64° 51′, to find the angle C and the sides AC and BC.

$$AC = 643.49 \text{ yards.}$$
$$BC = 600.11 \quad \text{"}$$

4. Find the height of D a point on a mountain above a horizontal plane. The angle of elevation at B, a point at the foot of the mountain, is 27° 29′; and at A distant from B 975 yards, in a direct line from B, and in the plane DBA, is 15° 36′.

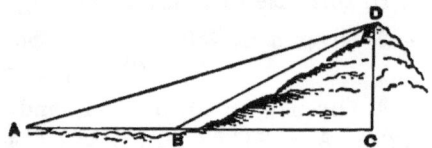

$$DC = 587.61 \text{ yards.}$$

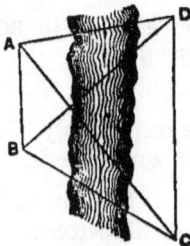

5. Wishing to know the distance between two inaccessible objects C and D, I measured a line, AB = 339 feet, from both ends of which the objects were visible; I found the angles BAD = 100°, BAC = 36° 30′, ABC = 121°, and ABD = 49°; find the distance DC.

$$DC = 697\frac{1}{2} \text{ feet.}$$

6. Wishing to know the distance between two inaccessible objects, A and B, and finding no place from which both could be seen, two points C and D, 200 yards distant, were found; from the former point A could be seen, and from the latter B; from C, a distance of 200 yards were measured to a point F, from which A could be seen; and from D the same distance was measured to E, from which B could be seen, and the following angles taken, viz.,

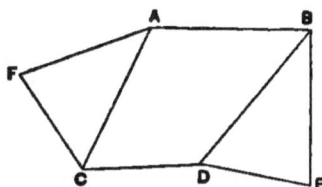

ACD = 53° 20′,	BDC = 156° 25′,
ACF = 54° 31′,	BDE = 54° 30′,
AFC = 83° 00′,	BED = 88° 30′.

Find the distance AB.   *Ans.* AB = 345.467 yards.

7. The distance between three points A, B, and C, are known, viz.,

AB = 800 yds.,

AC = 600 "

and   BC = 400 "

All are visible from a distant point P, at which the angles are measured,

APC = 33° 45′,   and   BPC = 22° 30′.

Find AP, BP, and CP.

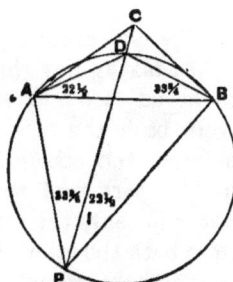

Ans. AP = 710.193 yds.,
        BP = 934.291 "
        CP = 1042.522 "

SPHERICAL TRIGONOMETRY.

Spherical Trigonometry treats of spherical triangles, the sides of which are arcs of great circles, each less than 180°, and the angles are diedral angles, formed by the planes of the great circles; each angle is less than two right angles.

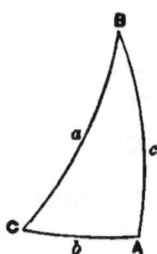

Napier's Five Circular Parts form the basis for the analysis of the functions of right-angled spherical triangles.

The two sides about the right angle, and the complements of the hypothenuse and of the two oblique angles are the five circular parts.

The spherical triangle ABC is right-angled at A. The sides b and c, and the complements of the hypothenuse a and of the angles B and C are the five circular parts.

In taking any three of these parts, they will either be found to be adjacent to each other, or one of them will be separated from both the others. When they are adjacent, the one lying between the others is called the middle part, and when they are not adjacent, the one separated from both the others is the middle part and the others are opposite.

Let ABC be a spherical triangle, right-angled at A, O the center of the sphere. Draw CD perpendicular to OA, and DE perpendicular to OB, and join CE. As the angle A is a right angle, the angle CDE is also a right angle, as CD is perpendicular to the

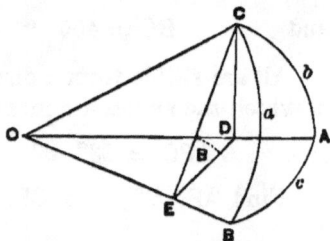

plane ABO in which DE is drawn perpendicular to OB, a line of the plane; hence CE is perpendicular to OB (Th. 3, Bk. 6), and CED = B.

1. $\sin b = \sin a \times \sin B = \cos \text{comp. } a \times \cos \text{comp. } B.$

2. $\sin c = \sin a \times \sin C = \cos \text{comp. } a \times \cos \text{comp. } C.$

3. $\cos B = \cos b \times \sin C = \cos b \times \cos \text{comp. } C = \sin \text{comp. } B.$

4. $\cos C = \cos c \times \sin B = \cos c \times \cos \text{comp. } B = \sin \text{comp. } C.$

5. $\cos a = \cos b \times \cos c = \cos b \times \cos c = \sin \text{comp. } a.$

By Prob. 4, in the triangle CED,

$$CD = CE \times \sin B,$$

$$\therefore \quad \sin b = \sin a \times \sin B. \tag{1}$$

No. 2 is derived in the same way, by making B the vertex instead of C.

$$\cos B = \frac{DE}{CE} = \frac{DE}{\sin a}.$$

$$DE = \cos b \times \sin a.$$

$$\sin c = \sin a \times \sin C. \tag{2}$$

In the triangle OED,

$$DE = OD \times \sin DOE,$$

$$\therefore \quad \cos B = \frac{\cos b \times \sin a \times \sin C}{\sin a} = \cos B \times \sin C. \tag{3}$$

In No. 4, cos C is found like cos B, by making B the vertex.

In the triangle ODE,

$$OE = OD \times \cos DOE;$$

that is, $\quad\quad \cos a = \cos b \times \cos c. \tag{5}$

From these five formulas, five others may be derived ; thus,

1. $\sin b = \sin a \times \sin B = \dfrac{\sin c \times \cos C}{\sin C \times \cos c} = \dfrac{\sin c}{\cos c} \times \dfrac{\cos C}{\sin C}$
$\quad\quad = \text{tang } c \times \cot C = \sin b.$

2. $\sin c = \sin a \times \sin C = \dfrac{\sin b \times \cos B}{\sin B \times \cos b} = \dfrac{\sin b}{\cos b} \times \dfrac{\cos B}{\sin B}$
$\quad\quad = \text{tang } b \times \cot B = \sin c.$

3. $\cos B = \cos b \times \sin C = \dfrac{\cos a \times \sin c}{\cos c \times \sin a} = \dfrac{\cos a}{\sin a} \times \dfrac{\sin c}{\cos c}$

$\qquad\qquad = \text{tang } c \times \cot a = \sin \text{ comp. B.}$

4. $\cos C = \cos c \times \sin B = \dfrac{\cos a \times \sin b}{\cos b \times \sin a} = \dfrac{\sin b}{\cos b} \times \dfrac{\cos a}{\sin a}$

$\qquad\qquad = \text{tang } b \times \cot a = \sin \text{ comp. C.}$

5. $\cos a = \cos b \times \cos c = \dfrac{\cos B \times \cos C}{\sin C \times \sin B} = \dfrac{\cos B}{\sin B} \times \dfrac{\cos C}{\sin C}$

$\qquad\qquad = \cot B \times \cot C = \sin \text{ comp. } a.$

From the first five formulas:
The sine of the middle part is equal to the product of the sines of the opposite parts.

From the second:
The sine of the middle part equals the product of the tangents of the adjacent parts.

REM.—Observe that the cosine of an angle is equal to the sine of the complement, and the cotangent is equal to the tangent of the complement.

THE SPECIES OF THE FUNCTIONS OF ANGLES OR ARCS.

As the functions of an arc and of its supplement are lines of equal length, there is a distinction necessary, in order that we may know whether the arc is greater or less than 90°; hence the minus sign is given to the cosine, the tangent, and the cotangent when the arc is greater than 90°, or terminates in the second quadrant.

Two arcs are said to be of the same species when they are both less or both greater than 90°, and of different species when the one is greater and the other less than 90°

1. From the 3d and 4th formulas of circular parts,

$$\sin C = \frac{\cos B}{\cos b}, \qquad \text{and} \qquad \sin B = \frac{\cos C}{\cos c}.$$

As the sines of C and B are both positive, hence the cosines of each oblique angle must have the same sign as the cosines of the opposite sides; consequently, the oblique angles and their opposite sides are of the same species.

2. When the hypothenuse is less than 90°, the other two sides and their opposite angles are of the same species; for, as

$$\cos a = \cos b \times \cos c,$$

and when a is less than 90° its cosine is positive; hence the cosines of b and c have like signs, that is, b and c are of the same species. But when a is greater than 90°, its cosine is negative; hence the cosines of b and c have different signs; that is, b and c are of different species.

By these two rules the nature of each result is determined, except when an oblique angle and the opposite side are given, to find the other parts.

Let ABC be right-angled at A; and B and b be known.

1st. If the sine of b is greater than the sine B, there can be no solution; for, as $\sin a = \dfrac{\sin b}{\sin B} > 1$, which is impossible.

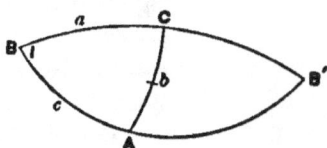

2d. If sine $b =$ sin B, then $\sin a = \dfrac{\sin b}{\sin B} = 1$; hence, the vertex B is the pole of the opposite side b, and a and c are each 90°.

3d. If sine b is less than sine B, when B is less than 90°, there will be two solutions, as shown in the above figure; as ABC and AB'C both fulfill the conditions. When B is greater than 90°; then, in order that sin $b <$ sin B, the side b must be greater than the angle B; when the result will be the same as above, and a and c in the one triangle will be complements of the same letters in the other triangle.

EXAMPLES.

1. Given $a = 86°\ 51'$, and $B = 18°\ 3'\ 32''$, to find b, c, and C.

  1. $\sin b\ = \sin a \times \sin B$,
  5. $\cos c\ = \cos a \div \cos b$;
  4. $\cos C = \cos c \times \sin B$.

$$\begin{aligned}
\log \sin B &= 18°\ \ 3'\ 32'' = 9.491354 \\
\cos c\ \ \ \ &= 86°\ 41'\ 14'' = \underline{8.761826} \\
C - 10\ &= 88°\ 58'\ 25'' = 8.253180 = \cos C.
\end{aligned}$$

$$\begin{aligned}
\log \sin a &= 86°\ 51'\ \ \ \ \ \ \ \ = 9.999343 \\
\log \sin b &= 18°\ \ 3'\ 32'' = 9.491354 \\
b - 10\ \ &= 18°\ \ 1'\ 50'' = \underline{9.490697} = \sin b.
\end{aligned}$$

$$\begin{aligned}
\log \cos a &= 86°\ 51'\ \ \ \ \ \ \ \ = 8.739969 \\
\log \cos b &= 18°\ \ 1'\ 50'' = 9.978143 \\
c + 10\ \ &= 86°\ 41'\ 14'' = \underline{9.761826} = \cos c.
\end{aligned}$$

2. Given $b = 155°\ 27'\ 54''$, and $c = 29°\ 46'\ 8''$, to find a, B, and C.
Ans. $a = 142°\ 9'\ 13''$, $B = 137°\ 24'\ 21''$, and $C = 54°\ 1'\ 16''$.

3. Given $B = 47°\ 13'\ 43''$, and $C = 126°\ 40'\ 24''$, to find a, b, and c.
  Ans. $a = 133°\ 32'\ 26''$, $b = 32°\ 8'\ 56''$, and $c = 144°\ 27'\ 3''$.

REM.—As the formulas are constructed with unity as radius, if logarithms are used, when the formula is a product, 10 must be subtracted, but when a quotient, 10 must be added.

A spherical triangle which has one of its sides a quadrant, is called a Quadrantal Triangle, and is readily solved by passing to its polar triangle, which will be right-angled, solving it, and returning to the quadrantal triangle.

The supplement of any side of a triangle is equal to the opposite angle of the polar triangle, and the supplement of any angle is equal to the opposite side of the polar triangle. The return is effected in the same way as each triangle is polar to the other.

PROBLEM I.

To show that the sines of the sides of a spherical triangle are respectively proportional to their opposite angles.

Let ABC be any oblique-angled triangle.

From either vertex, as A, draw an arc of a great circle perpendicular to the opposite side; then will the triangles ABD and ADE be right-angled at D, and

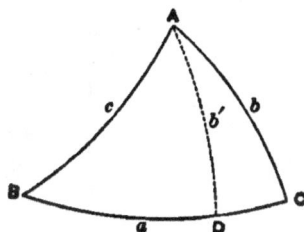

$$\sin b' = \sin c \times \sin B ;$$

and $$\sin b' = \sin b \times \sin C.$$

$$\therefore \quad \sin b \times \sin C = \sin c \times \sin B;$$

and $$\sin b : \sin c :: \sin B : \sin C.$$

In like manner, $\sin a : \sin b :: \sin A : \sin B$;

and $$\sin a : \sin c :: \sin A : \sin C.$$

The result is the same when the perpendicular falls on the opposite side produced.

In the triangle ABD and in the triangle ACD,

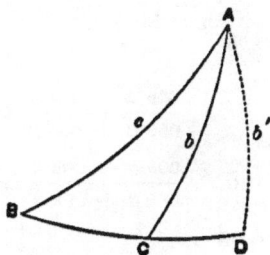

$$\sin b' = \sin c \times \sin B;$$

$$\sin b' = \sin b \times \sin C.$$

$$\therefore \quad \sin b \times \sin C = \sin c \times \sin B;$$

and

$$\sin b : \sin c :: \sin B : \sin C, \text{ etc.}$$

PROBLEM II.

In an oblique-angled spherical triangle, if from the vertex of either angle an arc be drawn perpendicular to the opposite side, dividing it into two segments, find these segments.

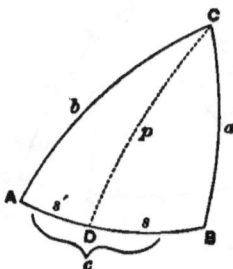

Let ABC be any oblique-angled spherical triangle.

From either vertex, as C, draw CD an arc of a great circle perpendicular to the opposite side; then, from 5th formula of Napier, $(s + s' = c)$.

In the triangle ACD and in the triangle BCD,

$$\cos b = \cos p \times \cos s';$$
$$\cos a = \cos p \times \cos s.$$

$$\therefore \quad \frac{\cos b}{\cos a} = \frac{\cos p \times \cos s'}{\cos p \times \cos s} = \frac{\cos s'}{\cos s},$$

and $\qquad \cos a : \cos b :: \cos s : \cos s';$

and by composition and division,

$$\cos a - \cos b : \cos a + \cos b :: \cos s - \cos s' : \cos s + \cos s'.$$

$$\therefore \quad \frac{\cos a - \cos b}{\cos a + \cos b} = \frac{\cos s - \cos s'}{\cos s + \cos s'};$$

and from Prob 5, Plane Trig.,

$$\frac{\cos M - \cos N}{\cos M + \cos N} = -\frac{\sin \frac{1}{2}(M + N) \times \sin \frac{1}{2}(M - N)}{\cos \frac{1}{2}(M + N) \times \cos \frac{1}{2}(M - N)}$$
$$= -\tan. \tfrac{1}{2}(M + N) \times \tan. \tfrac{1}{2}(M - N).$$

$$\therefore \quad \frac{\cos a - \cos b}{\cos a + \cos b} = -\tan. \tfrac{1}{2}(a + b) \times \tan. \tfrac{1}{2}(a - b);$$

and $\qquad \dfrac{\cos s - \cos s'}{\cos s + \cos s'} = -\tan. \tfrac{1}{2}(s + s') \times \tan. \tfrac{1}{2}(s - s').$

And

$$\tan. \tfrac{1}{2}(s + s') \times \tan. \tfrac{1}{2}(s - s') = \tan. \tfrac{1}{2}(a + b) \times \tan. \tfrac{1}{2}(a - b);$$
$$\therefore \quad \tan \tfrac{1}{2}(s + s') : \tan. \tfrac{1}{2}(a + b) :: \tan. \tfrac{1}{2}(a - b) : \tan. \tfrac{1}{2}(s - s').$$

REM.—s and s' being determined, in each right-angled triangle are known two sides and an angle opposite one of them.

PROBLEM III.

When two sides and the included angle are given, to find the other parts.

Let ABC be an oblique spherical triangle, a, c, and B given.

From A draw AD an arc of a great circle perpendicular to the opposite side BC, and in the triangle ABD,

$$\sin p = \sin c \times \sin B;$$

and

$$\sin BAD = \frac{\cos B}{\sin p};$$

and

$$\sin BD = \sin c \times \sin BAD.$$
$$DC = a - BD;$$
$$\cos b = \cos p \times \cos DC;$$
$$\sin C = \frac{\sin p}{\sin b};$$

and

$$\sin CAD = \frac{\sin DC}{\sin b}.$$

$$\text{angle } A = \text{angle } BAD + \text{angle } CAD;$$

hence, b, A, and C are determined.

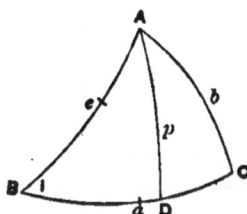

PROBLEM IV.

When a side and the two adjacent angles are given.

Let B, C, and a be given; then in the polar triangle,

$$b = 180 - B,$$
$$c = 180 - c,$$
$$\text{and} \quad A = 180 - a;$$

that is, two sides and the included angle known. Solve the polar triangle by Problem 3, and return to the original triangle.

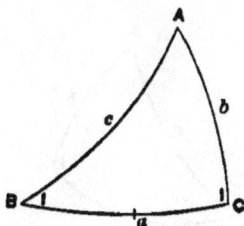

PROBLEM V.

When two sides and an angle opposite one of them is given.

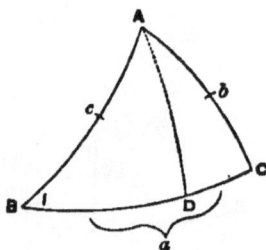

Let B, b, and c be given, and from the angle A, opposite the unknown side a, draw an arc of a great circle perpendicular to it.

In the triangle ABD,

$$\sin p = \sin c \times \sin B,$$

$$\sin BAD = \frac{\cos B}{\cos p},$$

and $\sin BD = \sin c \times \sin BAD.$

In the triangle ADC,

$$\cos DC = \frac{\cos b}{\cos p},$$

and $$\sin CAD = \frac{\sin DC}{\sin b},$$

angle A = angle BAD + angle CAD;

hence all the parts are determined.

REM.—When the three sides are given, the angles are found by this problem, after having found the segments of one side by Problem 2.

PROBLEM VI.

When two angles and a side opposite one of them is given.

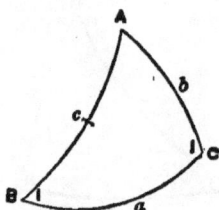

Let B, C, and c be given; then, in the polar triangle,

$$b = 180 - B,$$
$$c = 180 - C,$$

and $C = 180 - c;$

the same as in Problem 5, and must be solved accordingly, and then return to the original triangle.

PROBLEM VII.

To find the area of a spherical polygon.

When the angles are not given, find them by the foregoing problems; then from Geometry, Book 8,

The area of a spherical triangle is equal to the product of its spherical excess and the trirectangular triangle, and the same for any polygon.

EXAMPLES.

1. What is the area of a spherical triangle on the surface of a sphere whose diameter is 20 feet; the angles of the triangle are A = 130°; B = 110°; and C = 165°.

$$\begin{array}{l} 130 \\ 110 \\ 165 \\ \hline 405 \\ 180 \\ \hline 225 \end{array}$$

surface of trirectangular triangle $= \dfrac{20^2 \times 3.1416}{8}$

$= 157.08$

$157.08 \times \frac{1}{2} = 392.7$ sq. ft., *Ans.*

$\dfrac{225}{90} = \frac{1}{2}$, spherical excess.

2. What is the area of a spherical polygon of five sides on a sphere whose diameter is 40 feet, and the sum of the angles of the polygon is 660°.

$$\frac{40^2 \times 3.1416}{8} \times \frac{4}{3} = 40 \times 5 \times 1.0472$$

$$= 209.44 \text{ sq. ft., } Ans.$$

3. Find the area of a spherical polygon of eight sides, on a sphere 30 feet in diameter, and each angle of the polygon being 150 degrees.

$$\frac{30^2 \times 3.1416}{8} \times \frac{4}{3} = 30 \times 10 \times 1.5708$$

$$= 471.24 \text{ sq. ft., } Ans.$$

PROBLEM VIII.

To find the shortest distance, on the surface of the earth, between two places whose latitudes and longitudes are known.

Rem.—The shortest distance between two points on the surface of the earth is measured on the arc of a great circle joining the points.

EXAMPLES.

1. The latitude of New York City is 40° 48′; its longitude 3° east; the latitude of San Francisco is 37° 45′ north, and its longitude 45° 40′ west. What is the distance between them?

The radius of the earth is 3962 miles, making 69.15 miles to a degree. *Ans.* 37° 18′ 46″ = 2580.18 miles.

Let this figure represent a hemisphere; NS a meridian passing through Washington; EQ, equator. The point C represents New York, and B′ San Francisco; the point B is at the North Pole; BC and BB′ are the colatitudes of New York and San Francisco, and the angle B the difference of longitude of C and B′.

From C draw CA perpendicular to BB′; then in the triangle BB′C, angle B = 48° 40′, the side a = 49° 12′, and BB′ = 52° 15′;

and as $\quad \sin b = \sin a \times \sin B = 34° 38′ 23″$

and $\qquad \cos c = \dfrac{\cos a}{\cos b} = \qquad 37° 25′ 14″$

$\qquad c′ = 52° 15′ - 37° 25′ 14″ = 14° 49′ 46″;$

and $\quad \cos a′ = \cos b \times \cos c′ = 37° 18′ 46″ = 2580.18$ miles.

2. The latitude and longitude of New York given, also the distance from New York to San Francisco, and the latitude of the latter place, to find its longitude.

3. Given the latitude and longitude of New York, the distance to San Francisco and its longitude, to find the latitude.

Rem.—The student will become more familiar with the principles by finding the different parts of the same problem, than by taking different ones.

PROBLEM IX.

To find the hour of the day ; the altitude of the sun, its declination and the latitude of the observer being given.

The spherical triangle of which we know the three sides are in the celestial concave. Its vertices are the sun, the zenith of the observer, and the Celestial Pole, or the point in the heavens pierced by the axis of the earth, perpendicular to the equator.

The arc of the great circle joining the sun and the pole is the codeclination of the sun, when the sun and the observer are both on the same side of the equator; when they are on different sides of the equator, it is the sum of the declination and 90°.

The **Coaltitude** of the sun is the arc of the great circle joining the sun and the zenith of the observer; and the

Colatitude of the observer is the arc joining the zenith and the pole.

EXAMPLE.

In latitude 36° 40′ the declination of the sun is 12° 20′ N., and its altitude 30° 30′.   What is the hour of the day ?

Ans. Either 7 h. 56 m.   2 sec. A.M.,
or 4 h.   3 m. 58 sec. P.M.

In this example the three sides are given to find the angle at the pole, which is the hour angle, and being reduced from degrees, etc., to hours, minutes, etc., by dividing by 15, gives either the time before or after 12 M.   The angle having its vertex at the pole, one side of which extends from the pole to the sun, the other to the zenith.

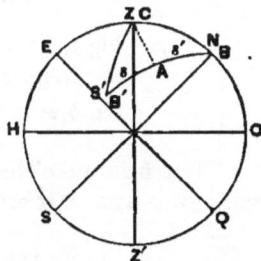

The sun, the zenith, and the celestial pole N are the vertices of the triangle BCB′; the three sides are given.   Draw CA perpendicular to BB′; then find the segments of c, s, and s'; and then the angle B, which reduce to hours, etc., and it is either so long before or after 12 o'clock M.

PROBLEM X.

To find the length of the day at any place, the latitude and declination of the sun being known.

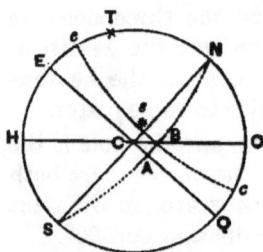

Let NS be the meridian at which the sun reaches the horizon when it is on the equator; that is, when it rises at 6 o'clock. When the sun has a declination north, it will be at *s* on the ecliptic, instead of being at C on the same meridian at 6 o'clock.

It has already passed the distance B*s* above the horizon, and the time taken for this passage is in the same proportion to 24 hours, that this arc AC is to 360 degrees. The angle is ANC, and is measured by the arc AC.

In the triangle ABC, right angled at A, AB is equal to the declination of the sun; the angle ACB = ECH is the coaltitude of the place.

EXAMPLE.

What is the length of the day in latitude 40° 30′ north, when the declination of the sun is 12° 50′?

T will be the position of the traveler.

$$\sin b = \cot C \times \text{tang. } c;$$

$$\therefore \quad
\begin{array}{lll}
\text{log tang. comp. C} & 40° 30' & = 9.931499 \\
\text{log tang. } c - 10 & 12° 50' & = 9.357566 \\
\sin b = \sin C & 11° 13' 10'' & 9.289065
\end{array}$$

Time before 6 o'clock that the sun rises and of course the same time after 6 it sets.

	h.	min.	sec.
11° 13′ 10″ =	0	44	53
			2
Hence,	1	29	46
	12	0	0
Length of day,	13	29	46

Twice the time of the sun passing from the horizon to the meridian NS must be added to 12 hours to get the length of the day.

REM.—As a traveler goes north, starting at the equator, for every degree that he travels, the south pole recedes one degree; therefore, the angle HCS measures his latitude, and HCE is his colatitude. This is the same as the north pole rising a degree for every degree he travels; hence, the altitude of the north pole is his latitude.

A

TABLE,

CONTAINING

THE LOGARITHMS OF NUMBERS

FROM 1 TO 10,000.

NUMBERS FROM 1 TO 100 AND THEIR LOGARITHMS,

WITH THEIR INDICES.

No.	Logarithm.	No.	Logarithm.	No.	Logarithm.	No.	Logarithm.	No.	Logarithm.
1	0·000000	21	1·322210	41	1·612784	61	1·785330	81	1·908485
2	0·301030	22	1·342423	42	1·623249	62	1·792392	82	1·913814
3	0·477121	23	1·361728	43	1·633468	63	1·799341	83	1·919078
4	0·602060	24	1·380211	44	1·643453	64	1·806180	84	1·924279
5	0·698970	25	1·397940	45	1·653213	65	1·812913	85	1·929419
6	0·778151	26	1·414973	46	1·662758	66	1·819544	86	1·934498
7	0·845098	27	1·431364	47	1·672098	67	1·826075	87	1·939519
8	0·903090	28	1·447158	48	1·681241	68	1·832500	88	1·944483
9	0·954243	29	1·462398	49	1·690196	69	1·838849	89	1·949390
10	1·000000	30	1·477121	50	1·698970	70	1·845098	90	1·954243
11	1·041393	31	1·491362	51	1·707570	71	1·851258	91	1·959041
12	1·079181	32	1·505150	52	1·716003	72	1·857332	92	1·963788
13	1·113943	33	1·518514	53	1·724276	73	1·863323	93	1·968483
14	1·146128	34	1·531479	54	1·732304	74	1·860232	94	1·973128
15	1·176001	35	1·544068	55	1·740363	75	1.875061	95	1·977724
16	1·204120	36	1·556303	56	1·748188	76	1·880814	96	1·982271
17	1·230449	37	1·568202	57	1·755875	77	1·886491	97	1·986772
18	1·255273	38	1·579784	58	1·763428	78	1·892095	98	1·991226
19	1·278754	39	1·591065	59	1·770852	79	1·897627	99	1·995635
20	1·301030	40	1·602060	60	1·778151	80	1·903090	100	2·000000

NOTE.—In the following part of the Table, the Indices are omitted, as they can be very easily supplied. Thus, the index of the logarithm of every integer number, consisting only of one number, is 0; of two figures, 1; of three figures, 2; of four figures, 3: being always a unit less than the number of figures contained in the integer number. The index to the logarithm of every number above 100, in the following part of the Table, is omitted; yet, in the operation, it must be prefixed, according to this remark: so that the logarithm of 600 is 2·77815, and that of 6000 is 3·77815, and so of the rest.

No.	0	1	2	3	4	5	6	7	8	9	Diff.
100	000000	000434	000868	001301	001734	002166	002598	003029	003461	003891	432
1	4321	4751	5181	5609	6038	6466	6894	7321	7748	8174	428
2	8600	9026	9451	9876	010300	010724	011147	011570	011993	012415	424
3	012837	013259	013680	014100	4521	4940	5360	5779	6197	6616	420
4	7033	7451	7868	8284	8700	9116	9532	9947	020361	020775	416
5	021189	021603	022016	022428	022841	023252	023664	024075	4486	4896	412
6	5306	5715	6125	6533	6942	7350	7757	8164	8571	8978	408
7	9384	9789	030195	030600	031004	031408	031812	032216	032619	033021	404
8	033424	033826	4227	4628	5029	5430	5830	6230	6629	7028	400
9	7426	7825	8223	8620	9017	9414	9811	040207	040602	040998	397
110	041393	041787	042182	042576	042969	043362	043755	044148	044540	044932	393
1	5323	5714	6105	6495	6885	7275	7664	8053	8442	8830	390
2	9218	9606	9993	050380	050766	051153	051538	051924	052309	052694	386
3	053078	053463	053846	4230	4613	4996	5378	5760	6142	6524	383
4	6905	7286	7666	8046	8426	8805	9185	9563	9942	060320	379
5	060698	061075	061452	061829	062206	062582	062958	063333	063709	4083	376
6	4458	4832	5206	5580	5953	6326	6699	7071	7443	7815	373
7	8186	8557	8928	9298	9668	070038	070407	070776	071145	071514	370
8	071882	072250	072617	072985	073352	3718	4085	4451	4816	5182	366
9	5547	5912	6276	6640	7004	7368	7731	8094	8457	8819	363
120	079181	070543	079904	080266	080626	080987	081347	081707	082067	082426	360
1	082785	083144	083503	3861	4219	4576	4934	5291	5647	6004	357
2	6360	6716	7071	7426	7781	8136	8490	8845	9198	9552	355
3	9905	090258	090611	090963	091315	091667	092018	092370	092721	093071	352
4	093422	3772	4122	4471	4820	5169	5518	5866	6215	6562	349
5	6910	7257	7604	7951	8298	8644	8990	9335	9681	100026	346
6	100371	100715	101059	101403	101747	102091	102434	102777	103119	3462	343
7	3804	4146	4487	4828	5169	5510	5851	6191	6531	6871	341
8	7210	7549	7888	8227	8565	8903	9241	9579	9916	110253	338
9	110590	110926	111263	111599	111934	112270	112605	112940	113275	3609	335
130	113943	114277	114611	114944	115278	115611	115943	116276	116608	116940	333
1	7271	7603	7934	8265	8595	8926	9256	9586	9915	120245	330
2	120574	120903	121231	121560	121888	122216	122544	122871	123198	3525	328
3	3852	4178	4504	4830	5156	5481	5806	6131	6456	6781	325
4	7105	7429	7753	8076	8399	8722	9045	9368	9690	130012	323
5	130334	130655	130977	131298	131619	131939	132260	132580	132900	3219	321
6	3539	3858	4177	4496	4814	5133	5451	5769	6086	6403	318
7	6721	7037	7354	7671	7987	8303	8618	8934	9249	9564	316
8	9879	140194	140508	140822	141136	141450	141763	142076	142389	142702	314
9	143015	3327	3639	3951	4263	4574	4885	5196	5507	5818	311
140	146128	146438	146748	147058	147367	147676	147985	148294	148603	148911	309
1	9219	9527	9835	150142	150449	150756	151063	151370	151676	151982	307
2	152288	152594	152900	3205	3510	3815	4120	4424	4728	5032	305
3	5336	5640	5943	6246	6549	6852	7154	7457	7759	8061	303
4	8362	8664	8965	9266	9567	9868	160168	160469	160769	161068	301
5	161368	161667	161967	162266	162564	162863	3161	3460	3758	4055	299
6	4353	4650	4947	5244	5541	5838	6134	6430	6726	7022	297
7	7317	7613	7908	8203	8497	8792	9086	9380	9674	9968	295
8	170262	170555	170848	171141	171434	171726	172019	172311	172603	172895	293
9	3186	3478	3769	4060	4351	4641	4932	5222	5512	5802	291
150	176091	176381	176670	176959	177248	177536	177825	178113	178401	178689	289
1	8977	9264	9552	9839	180126	180413	180699	180986	181272	181558	287
2	181844	182129	182415	182700	2985	3270	3555	3839	4123	4407	285
3	4691	4975	5259	5542	5825	6108	6391	6674	6956	7239	283
4	7521	7803	8084	8366	8647	8928	9209	9490	9771	190051	281
5	190332	190612	190892	191171	191451	191730	192010	192289	192567	2846	279
6	3125	3403	3681	3959	4237	4514	4792	5069	5346	5623	278
7	5900	6176	6453	6729	7005	7281	7556	7832	8107	8382	276
8	8657	8932	9206	9481	9755	200029	200303	200577	200850	201124	274
9	201397	201670	201943	202216	202488	2761	3033	3305	3577	3848	272

| No. | 0 | 1 | 2 | 3 | 4 | 5 | 6 | 7 | 8 | 9 | Diff. |

No.	0	1	2	3	4	5	6	7	8	9	Diff
160	204120	204391	204663	204934	205204	205475	205746	206016	206286	206550	271
1	6826	7096	7365	7634	7904	8173	8441	8710	8979	9247	269
2	9515	9783	210051	210319	210586	210853	211121	211388	211654	211921	267
3	212188	212454	2720	2986	3252	3518	3783	4049	4314	4579	266
4	4844	5109	5373	5638	5902	6166	6430	6694	6957	7221	264
5	7484	7747	8010	8273	8536	8798	9000	9323	9585	9846	262
6	220108	220370	220631	220892	221153	221414	221675	221936	222196	222456	261
7	2716	2976	3236	3496	3755	4015	4274	4533	4792	5051	259
8	5309	5568	5826	6084	6342	6600	6858	7115	7372	7630	258
9	7887	8144	8400	8657	8913	9170	9426	9682	9938	230193	256
170	230449	230704	230960	231215	231470	231724	231979	232234	232488	232742	255
1	2996	3250	3504	3757	4011	4264	4517	4770	5023	5276	253
2	5528	5781	6033	6285	6537	6789	7041	7292	7544	7795	252
3	8046	8297	8548	8799	9049	9299	9550	9800	240050	240300	250
4	240549	240799	241048	241297	241546	241795	242044	242293	2541	2790	249
5	3038	3286	3534	3782	4030	4277	4525	4772	5019	5266	248
6	5513	5759	6006	6252	6499	6745	6991	7237	7482	7728	246
7	7973	8219	8464	8709	8954	9198	9443	9687	9932	250176	245
8	250420	250664	250908	251151	251395	251638	251881	252125	252368	2610	243
9	2853	3096	3338	3580	3822	4064	4306	4548	4790	5031	242
180	255273	255514	255755	255996	256237	256477	256718	256958	257198	257439	241
1	7679	7918	8158	8398	8637	8877	9116	9355	9594	9833	239
2	260071	260310	260548	260787	261025	261263	261501	261739	261976	262214	238
3	2451	2688	2925	3162	3399	3636	3873	4109	4346	4582	237
4	4818	5054	5290	5525	5761	5996	6232	6467	6702	6937	235
5	7172	7406	7641	7875	8110	8344	8578	8812	9046	9279	234
6	9513	9746	9980	270213	270446	270679	270912	271144	271377	271609	233
7	271842	272074	272306	2538	2770	3001	3233	3464	3696	3927	232
8	4158	4389	4620	4850	5081	5311	5542	5772	6002	6232	230
9	6462	6692	6921	7151	7380	7609	7838	8067	8296	8525	229
190	278754	278982	279211	279439	279667	279895	280123	280351	280578	280806	228
1	281033	281261	281488	281715	281942	282169	282396	2622	2849	3075	227
2	3301	3527	3753	3979	4205	4431	4656	4882	5107	5332	226
3	5557	5782	6007	6232	6456	6681	6905	7130	7354	7578	225
4	7802	8026	8249	8473	8696	8920	9143	9366	9589	9812	223
5	290035	290257	290480	290702	290925	291147	291369	291591	291813	292034	222
6	2256	2478	2699	2920	3141	3363	3584	3804	4025	4246	221
7	4466	4687	4907	5127	5347	5567	5787	6007	6226	6446	220
8	6665	6884	7104	7323	7542	7761	7979	8198	8416	8635	219
9	8853	9071	9289	9507	9725	9943	300161	300378	300595	300813	218
200	301030	301247	301464	301681	301898	302114	302331	302547	302764	302980	217
1	3196	3412	3628	3844	4059	4275	4491	4706	4921	5136	216
2	5351	5566	5781	5996	6211	6425	6639	6854	7068	7282	215
3	7496	7710	7924	8137	8351	8564	8778	8991	9204	9417	213
4	9630	9843	310056	310268	310481	310693	310906	311118	311330	311542	212
5	311754	311966	2177	2389	2600	2812	3023	3234	3445	3656	211
6	3867	4078	4289	4499	4710	4920	5130	5340	5551	5760	210
7	5970	6180	6390	6599	6809	7018	7227	7436	7646	7854	209
8	8063	8272	8481	8689	8898	9106	9314	9522	9730	9938	208
9	320146	320354	320562	320769	320977	321184	321391	321598	321805	322012	207
210	322219	322426	322633	322839	323046	323252	323458	323665	323871	324077	206
1	4282	4488	4694	4899	5105	5310	5516	5721	5926	6131	205
2	6336	6541	6745	6950	7155	7359	7563	7767	7972	8176	204
3	8380	8583	8787	8991	9194	9398	9601	9805	330008	330211	203
4	330414	330617	330819	331022	331225	331427	331630	331832	2034	2236	202
5	2438	2640	2842	3044	3246	3447	3649	3850	4051	4253	202
6	4454	4655	4856	5057	5257	5458	5658	5859	6059	6260	201
7	6460	6660	6860	7060	7260	7459	7659	7858	8058	8257	200
8	8456	8656	8855	9054	9253	9451	9650	9849	340046	340246	199
9	340444	340642	340841	341039	341237	341435	341632	341830	2028	2225	198

| No. | 0 | 1 | 2 | 3 | 4 | 5 | 6 | 7 | 8 | 9 | Diff |

No.	0	1	2	3	4	5	6	7	8	9	Diff.
220	342423	342620	342817	343014	343212	343409	343606	343802	343999	344196	197
1	4392	4589	4785	4981	5178	5374	5570	5766	5962	6157	196
2	6353	6549	6744	6939	7135	7330	7525	7720	7915	8110	195
3	8305	8500	8694	8889	9083	9278	9472	9666	9860	350054	194
4	350248	350442	350636	350829	351023	351216	351410	351603	351796	1989	193
5	2183	2375	2568	2761	2954	3147	3339	3532	3724	3916	193
6	4108	4301	4493	4685	4876	5068	5260	5452	5643	5834	192
7	6026	6217	6408	6599	6790	6981	7172	7363	7554	7744	191
8	7935	8125	8316	8506	8696	8886	9076	9266	9456	9646	190
9	9835	360025	360215	360404	360593	360783	360972	361161	361350	361539	189
230	361728	361917	362105	362294	362482	362671	362859	363048	363236	363424	188
1	3612	3800	3988	4176	4363	4551	4739	4926	5113	5301	188
2	5488	5675	5862	6049	6236	6423	6610	6796	6983	7169	187
3	7356	7542	7729	7915	8101	8287	8473	8659	8845	9030	186
4	9216	9401	9587	9772	9958	370143	370328	370513	370698	370883	185
5	371068	371253	371437	371622	371806	1991	2175	2360	2544	2728	184
6	2912	3096	3280	3464	3647	3831	4015	4198	4382	4565	184
7	4748	4932	5115	5298	5481	5664	5846	6029	6212	6394	183
8	6577	6759	6942	7124	7306	7488	7670	7852	8034	8216	182
9	8398	8580	8761	8943	9124	9306	9487	9668	9849	380030	181
240	380211	380392	380573	380754	380934	381115	381296	381476	381656	381837	181
1	2017	2197	2377	2557	2737	2917	3097	3277	3456	3636	180
2	3815	3995	4174	4353	4533	4712	4891	5070	5249	5428	179
3	5606	5785	5964	6142	6321	6499	6677	6856	7034	7212	178
4	7390	7568	7746	7923	8101	8279	8456	8634	8811	8989	178
5	9166	9343	9520	9698	9875	390051	390228	390405	390582	390759	177
6	390935	391112	391288	391464	391641	1817	1993	2169	2345	2521	176
7	2697	2873	3048	3224	3400	3575	3751	3926	4101	4277	176
8	4452	4627	4802	4977	5152	5326	5501	5676	5850	6025	175
9	6199	6374	6548	6722	6896	7071	7245	7419	7592	7766	174
250	397940	398114	398287	398461	398634	398808	398981	399154	399328	399501	173
1	9674	9847	400020	400192	400365	400538	400711	400883	401056	401228	173
2	401401	401573	1745	1917	2089	2261	2433	2605	2777	2949	172
3	3121	3292	3464	3635	3807	3978	4149	4320	4492	4663	171
4	4834	5005	5176	5346	5517	5688	5858	6029	6199	6370	171
5	6540	6710	6881	7051	7221	7391	7561	7731	7901	8070	170
6	8240	8410	8579	8749	8918	9087	9257	9426	9595	9764	169
7	9933	410102	410271	410440	410609	410777	410946	411114	411283	411451	169
8	411620	1788	1956	2124	2293	2461	2629	2796	2964	3132	168
9	3300	3467	3635	3803	3970	4137	4305	4472	4639	4806	167
260	414973	415140	415307	415474	415641	415808	415974	416141	416308	416474	167
1	6641	6807	6973	7139	7306	7472	7638	7804	7970	8135	166
2	8301	8467	8633	8798	8964	9129	9295	9460	9625	9791	165
3	9956	420121	420286	420451	420616	420781	420945	421110	421275	421439	165
4	421604	1768	1933	2097	2261	2426	2590	2754	2918	3082	164
5	3246	3410	3574	3737	3901	4065	4228	4392	4555	4718	164
6	4882	5045	5208	5371	5534	5697	5860	6023	6186	6349	163
7	6511	6674	6836	6999	7161	7324	7486	7648	7811	7973	162
8	8135	8297	8459	8621	8783	8944	9106	9268	9429	9591	162
9	9752	9914	430075	430236	430398	430559	430720	430881	431042	431203	161
270	431364	431525	431685	431846	432007	432167	432328	432488	432649	432809	161
1	2960	3130	3290	3450	3610	3770	3930	4090	4249	4409	160
2	4569	4729	4888	5048	5207	5367	5526	5685	5844	6004	159
3	6163	6322	6481	6640	6799	6957	7116	7275	7433	7592	159
4	7751	7909	8067	8226	8384	8542	8701	8859	9017	9175	158
5	9333	9491	9648	9806	9964	440122	440279	440437	440594	440752	158
6	440909	441066	441224	441381	441538	1695	1852	2009	2166	2323	157
7	2480	2637	2793	2950	3106	3263	3419	3576	3732	3889	157
8	4045	4201	4357	4513	4669	4825	4981	5137	5293	5449	156
9	5604	5760	5915	6071	6226	6382	6537	6692	6848	7003	155

| No. | 0 | 1 | 2 | 3 | 4 | 5 | 6 | 7 | 8 | 9 | Diff. |

No.	0	1	2	3	4	5	6	7	8	9	Diff.
280	447158	447313	447468	447623	447778	447933	448088	448242	448397	448552	155
1	8706	8861	9015	9170	9324	9478	9633	9787	9941	450095	154
2	450249	450403	450557	450711	450865	451018	451172	451326	451479	1633	154
3	1786	1940	2093	2247	2400	2553	2706	2859	3012	3165	153
4	3318	3471	3624	3777	3930	4082	4235	4387	4540	4692	153
5	4845	4997	5150	5302	5454	5606	5758	5910	6062	6214	152
6	6306	6518	6670	6821	6973	7125	7276	7428	7579	7731	152
7	7882	8033	8184	8336	8487	8638	8789	8940	9091	9242	151
8	9392	9543	9694	9845	9995	460146	460296	460447	460597	460748	151
9	460898	461048	461198	461348	461499	1649	1799	1948	2098	2248	150
290	462398	462548	462697	462847	462997	463146	463296	463445	463594	463744	150
1	3893	4042	4191	4340	4490	4639	4788	4936	5085	5234	149
2	5383	5532	5680	5829	5977	6126	6274	6423	6571	6719	149
3	6868	7016	7164	7312	7460	7608	7756	7904	8052	8200	148
4	8347	8495	8643	8790	8938	9085	9233	9380	9527	9675	148
5	9822	9969	470116	470263	470410	470557	470704	470851	470998	471145	147
6	471292	471438	1585	1732	1878	2025	2171	2318	2464	2610	146
7	2756	2903	3049	3195	3341	3487	3633	3779	3925	4071	146
8	4216	4362	4508	4653	4799	4944	5090	5235	5381	5526	146
9	5671	5816	5962	6107	6252	6397	6542	6687	6832	6976	145
300	477121	477266	477411	477555	477700	477844	477989	478133	478278	478422	145
1	8560	8711	8855	8999	9143	9287	9431	9575	9719	9863	144
2	480007	480151	480294	480438	480582	480725	480869	481012	481156	481299	144
3	1443	1586	1729	1872	2016	2159	2302	2445	2588	2731	143
4	2874	3016	3159	3302	3445	3587	3730	3872	4015	4157	143
5	4300	4442	4585	4727	4869	5011	5153	5295	5437	5570	142
6	5721	5863	6005	6147	6289	6430	6572	6714	6855	6997	142
7	7138	7280	7421	7563	7704	7845	7986	8127	8269	8410	141
8	8551	8692	8833	8974	9114	9255	9396	9537	9677	9818	141
9	9958	490099	490239	490380	490520	490661	490801	490941	491081	491222	140
310	491362	491502	491642	491782	491922	492062	492201	492341	492481	492621	140
1	2760	2900	3040	3179	3319	3458	3597	3737	3876	4015	139
2	4155	4294	4433	4572	4711	4850	4989	5128	5267	5406	139
3	5544	5683	5822	5960	6099	6238	6376	6515	6653	6791	139
4	6930	7068	7206	7344	7483	7621	7759	7897	8035	8173	138
5	8311	8448	8586	8724	8862	8999	9137	9275	9412	9550	138
6	9687	9824	9962	500099	500236	500374	500511	500648	500785	500922	137
7	501059	501196	501333	1470	1607	1744	1880	2017	2154	2291	137
8	2427	2564	2700	2837	2973	3109	3246	3382	3518	3655	136
9	3791	3927	4063	4199	4335	4471	4607	4743	4878	5014	136
320	505150	505286	505421	505557	505693	505828	505964	506099	506234	506370	136
1	6505	6640	6776	6911	7046	7181	7316	7451	7586	7721	135
2	7856	7991	8126	8260	8395	8530	8664	8799	8934	9068	135
3	9203	9337	9471	9606	9740	9874	510009	510143	510277	510411	134
4	510545	510679	510813	510947	511081	511215	1349	1482	1616	1750	134
5	1883	2017	2151	2284	2418	2551	2684	2818	2951	3084	133
6	3218	3351	3484	3617	3750	3883	4016	4149	4282	4415	133
7	4548	4681	4813	4946	5079	5211	5344	5476	5609	5741	133
8	5874	6006	6139	6271	6403	6535	6668	6800	6932	7064	132
9	7196	7328	7460	7592	7724	7855	7987	8119	8251	8382	132
330	518514	518646	518777	518909	519040	519171	519303	519434	519566	519697	131
1	9828	9959	520090	520221	520353	520484	520615	520745	520876	521007	131
2	521138	521269	1400	1530	1661	1792	1922	2053	2183	2314	131
3	2444	2575	2705	2835	2966	3096	3226	3356	3486	3616	130
4	3746	3876	4006	4136	4266	4396	4526	4656	4785	4915	130
5	5045	5174	5304	5434	5563	5693	5822	5951	6081	6210	129
6	6339	6469	6598	6727	6856	6985	7114	7243	7372	7501	129
7	7630	7759	7888	8016	8145	8274	8402	8531	8660	8788	129
8	8917	9045	9174	9302	9430	9559	9687	9815	9943	530072	128
9	530200	530328	530456	530584	530712	530840	530968	531096	531223	1351	128
No.	0	1	2	3	4	5	6	7	8	9	Diff.

No.	0	1	2	3	4	5	6	7	8	9	Diff.
340	531479	531607	531734	531862	531990	532117	532245	532372	532500	532627	128
1	2754	2882	3009	3136	3264	3391	3518	3645	3772	3899	127
2	4026	4153	4280	4407	4534	4661	4787	4914	5041	5167	127
3	5294	5421	5547	5674	5800	5927	6053	6180	6306	6432	126
4	6558	6685	6811	6937	7063	7189	7315	7441	7567	7693	126
5	7819	7945	8071	8197	8322	8448	8574	8699	8825	8951	126
6	9076	9202	9327	9452	9578	9703	9829	9954	540079	540204	125
7	540329	540455	540580	540705	540830	540955	541080	541205	1330	1454	125
8	1579	1704	1829	1953	2078	2203	2327	2452	2576	2701	125
9	2825	2950	3074	3199	3323	3447	3571	3696	3820	3944	124
350	544068	544192	544316	544440	544564	544688	544812	544936	545060	545183	124
1	5307	5431	5555	5678	5802	5925	6049	6172	6296	6419	124
2	6543	6666	6789	6913	7036	7159	7282	7405	7529	7652	123
3	7775	7898	8021	8144	8267	8389	8512	8635	8758	8881	123
4	9003	9126	9240	9371	9494	9616	9739	9861	9984	550106	123
5	550228	550351	550473	550595	550717	550840	550962	551084	551206	1328	122
6	1450	1572	1694	1816	1938	2060	2181	2303	2425	2547	122
7	2668	2790	2911	3033	3155	3276	3398	3519	3640	3762	121
8	3883	4004	4126	4247	4368	4489	4610	4731	4852	4973	121
9	5094	5215	5336	5457	5578	5699	5820	5940	6061	6182	121
360	556303	556423	556544	556664	556785	556905	557026	557146	557267	557387	120
1	7507	7627	7748	7868	7988	8108	8228	8349	8469	8589	120
2	8709	8829	8948	9068	9188	9308	9428	9548	9667	9787	120
3	9907	560026	560146	560265	560385	560504	560624	560743	560863	560982	119
4	561101	1221	1340	1459	1578	1698	1817	1936	2055	2174	119
5	2293	2412	2531	2650	2769	2887	3006	3125	3244	3362	119
6	3481	3600	3718	3837	3955	4074	4192	4311	4429	4548	119
7	4666	4784	4903	5021	5139	5257	5376	5494	5612	5730	118
8	5848	5966	6084	6202	6320	6437	6555	6673	6791	6909	118
9	7026	7144	7262	7379	7497	7614	7732	7849	7967	8084	118
370	568202	568319	568436	568554	568671	568788	568905	569023	569140	569257	117
1	9374	9491	9608	9725	9842	9959	570076	570193	570309	570426	117
2	570543	570660	570776	570893	571010	571126	1243	1359	1476	1592	117
3	1709	1825	1942	2058	2174	2291	2407	2523	2639	2755	116
4	2872	2988	3104	3220	3336	3452	3568	3684	3800	3915	116
5	4031	4147	4263	4379	4494	4610	4726	4841	4957	5072	116
6	5198	5303	5419	5534	5650	5765	5880	5996	6111	6226	115
7	6341	6457	6572	6687	6802	6917	7032	7147	7262	7377	115
8	7492	7607	7722	7836	7951	8066	8181	8295	8410	8525	115
9	8639	8754	8868	8983	9097	9212	9326	9441	9555	9669	114
380	579784	579898	580012	580126	580241	580355	580469	580583	580697	580811	114
1	580925	581039	1153	1267	1381	1495	1608	1722	1836	1950	114
2	2063	2177	2291	2404	2518	2631	2745	2858	2972	3085	114
3	3199	3312	3426	3539	3652	3765	3879	3992	4105	4218	113
4	4331	4444	4557	4670	4783	4896	5009	5122	5235	5348	113
5	5461	5574	5686	5799	5912	6024	6137	6250	6362	6475	113
6	6587	6700	6812	6925	7037	7149	7262	7374	7486	7599	112
7	7711	7823	7935	8047	8160	8272	8384	8496	8608	8720	112
8	8832	8944	9056	9167	9279	9391	9503	9615	9726	9838	112
9	9950	590061	590173	590284	590396	590507	590619	590730	590842	590953	112
390	591065	591176	591287	591399	591510	591621	591732	591843	591955	592066	111
1	2177	2288	2399	2510	2621	2732	2843	2954	3064	3175	111
2	3286	3397	3508	3618	3729	3840	3950	4061	4171	4282	111
3	4393	4503	4614	4724	4834	4945	5055	5165	5276	5386	110
4	5496	5606	5717	5827	5937	6047	6157	6267	6377	6487	110
5	6597	6707	6817	6927	7037	7146	7256	7366	7476	7586	110
6	7695	7805	7914	8024	8134	8243	8353	8462	8572	8681	110
7	8791	8900	9009	9119	9228	9337	9446	9556	9665	9774	109
8	9883	9992	600101	600210	600319	600428	600537	600646	600755	600864	109
9	600973	601082	1191	1299	1408	1517	1625	1734	1843	1951	109
No.	0	1	2	3	4	5	6	7	8	9	Diff.

No.	0	1	2	3	4	5	6	7	8	9	Diff.
400	602060	602169	602277	602386	602494	602603	602711	602819	602928	603036	108
1	3144	3253	3361	3409	3577	3686	3794	3902	4010	4118	108
2	4226	4334	4442	4550	4658	4766	4874	4982	5089	5197	108
3	5305	5413	5521	5628	5736	5844	5951	6059	6166	6274	108
4	6381	6489	6596	6704	6811	6919	7026	7133	7241	7348	107
5	7455	7562	7669	7777	7884	7991	8098	8205	8312	8419	107
6	8526	8633	8740	8847	8954	9061	9167	9274	9381	9488	107
7	9594	9701	9808	9914	610021	610128	610234	610341	610447	610554	107
8	610660	610767	610873	610979	1080	1192	1298	1405	1511	1617	106
9	1723	1829	1936	2042	2148	2254	2360	2466	2572	2678	106
410	612784	612890	612996	613102	613207	613313	613419	613525	613630	613736	106
1	3842	3947	4053	4159	4264	4370	4475	4581	4686	4792	106
2	4897	5003	5108	5213	5319	5424	5529	5634	5740	5845	105
3	5950	6055	6160	6265	6370	6476	6581	6686	6790	6895	105
4	7000	7105	7210	7315	7420	7525	7629	7734	7830	7943	105
5	8048	8153	8257	8362	8466	8571	8676	8780	8884	8989	105
6	9093	9198	9302	9406	9511	9615	9719	9824	9928	620032	104
7	620136	620240	620344	620448	620552	620656	620760	620864	620968	1072	104
8	1176	1280	1384	1488	1592	1695	1799	1903	2007	2110	104
9	2214	2318	2421	2525	2628	2732	2835	2939	3042	3146	104
420	623249	623353	623456	623559	623663	623766	623869	623973	624076	624179	103
1	4282	4385	4488	4591	4695	4798	4901	5004	5107	5210	103
2	5312	5415	5518	5621	5724	5827	5929	6032	6135	6238	103
3	6340	6443	6546	6648	6751	6853	6956	7058	7161	7263	103
4	7366	7468	7571	7673	7775	7878	7980	8082	8185	8287	102
5	8389	8491	8593	8695	8797	8900	9002	9104	9206	9308	102
6	9410	9512	9613	9715	9817	9919	630021	630123	630224	630326	102
7	630428	630530	630631	630733	630835	630936	1038	1139	1241	1342	102
8	1444	1545	1647	1748	1849	1951	2052	2153	2255	2356	101
9	2457	2559	2660	2761	2862	2963	3064	3165	3266	3367	101
430	633468	633569	633670	633771	633872	633973	634074	634175	634276	634376	101
1	4477	4578	4679	4779	4880	4981	5081	5182	5283	5383	101
2	5484	5584	5685	5785	5886	5986	6087	6187	6287	6388	100
3	6488	6588	6688	6789	6889	6989	7089	7189	7290	7390	100
4	7490	7590	7690	7790	7890	7990	8090	8190	8290	8389	100
5	8489	8589	8689	8789	8888	8988	9088	9188	9287	9387	100
6	9486	9586	9686	9785	9885	9984	640084	640183	640283	640382	99
7	640481	640581	640680	640779	640879	640978	1077	1177	1276	1375	99
8	1474	1573	1672	1771	1871	1970	2069	2168	2267	2366	99
9	2465	2563	2662	2761	2860	2959	3058	3156	3255	3354	99
440	643453	643551	643650	643749	643847	643946	644044	644143	644242	644340	98
1	4439	4537	4636	4734	4832	4931	5029	5127	5226	5324	98
2	5422	5521	5619	5717	5815	5913	6011	6110	6208	6306	98
3	6404	6502	6600	6698	6796	6894	6992	7089	7187	7285	98
4	7383	7481	7579	7676	7774	7872	7969	8067	8165	8262	98
5	8360	8458	8555	8653	8750	8848	8945	9043	9140	9237	97
6	9335	9432	9530	9627	9724	9821	9919	650016	650113	650210	97
7	650308	650405	650502	650599	650696	650793	650890	0987	1084	1181	97
8	1278	1375	1472	1569	1666	1762	1859	1956	2053	2150	97
9	2246	2343	2440	2536	2633	2730	2826	2923	3019	3116	97
450	653213	653309	653405	653502	653598	653695	653791	653888	653984	654080	96
1	4177	4273	4369	4465	4562	4658	4754	4850	4946	5042	96
2	5138	5235	5331	5427	5523	5619	5715	5810	5906	6002	96
3	6098	6194	6290	6386	6482	6577	6673	6769	6864	6960	96
4	7056	7152	7247	7343	7438	7534	7629	7725	7820	7916	96
5	8011	8107	8202	8298	8393	8488	8584	8679	8774	8870	95
6	8965	9060	9155	9250	9346	9441	9536	9631	9726	9821	95
7	9916	660011	660106	660201	660296	660391	660486	660581	660676	660771	95
8	660865	0960	1055	1150	1245	1339	1434	1529	1623	1718	95
9	1813	1907	2002	2096	2191	2286	2380	2475	2569	2663	95
No.	0	1	2	3	4	5	6	7	8	9	Diff.

No.	0	1	2	3	4	5	6	7	8	9	Diff.
460	662758	662852	662947	663041	663135	663230	663324	663418	663512	663607	94
1	3701	3795	3889	3983	4078	4172	4266	4360	4454	4548	94
2	4642	4736	4830	4924	5018	5112	5206	5299	5393	5487	94
3	5581	5675	5769	5862	5956	6050	6143	6237	6331	6424	94
4	6518	6612	6705	6799	6892	6986	7079	7173	7266	7360	94
5	7453	7546	7640	7733	7826	7920	8013	8106	8199	8293	93
6	8386	8479	8572	8665	8759	8852	8945	9038	9131	9224	93
7	9317	9410	9503	9596	9689	9782	9875	9967	670060	670153	93
8	670246	670339	670431	670524	670617	670710	670802	670895	0988	1080	93
9	1173	1265	1358	1451	1543	1636	1728	1821	1913	2005	93
470	672098	672190	672283	672375	672467	672560	672652	672744	672836	672929	92
1	3021	3113	3205	3297	3390	3482	3574	3666	3758	3830	92
2	3942	4034	4126	4218	4310	4402	4494	4586	4677	4769	92
3	4861	4953	5045	5137	5228	5320	5412	5503	5595	5687	92
4	5778	5870	5962	6053	6145	6236	6328	6419	6511	6602	92
5	6694	6785	6876	6968	7059	7151	7242	7333	7424	7516	91
6	7607	7698	7789	7881	7972	8063	8154	8245	8336	8427	91
7	8518	8609	8700	8791	8882	8973	9064	9155	9246	9337	91
8	9428	9519	9610	9700	9791	9882	9973	680063	680154	680245	91
9	680336	680426	680517	680607	680698	680789	680879	0970	1060	1151	91
480	681241	681332	681422	681513	681603	681693	681784	681874	681964	682055	90
1	2145	2235	2326	2416	2506	2596	2686	2777	2867	2957	90
2	3047	3137	3227	3317	3407	3497	3587	3677	3767	3857	90
3	3947	4037	4127	4217	4307	4396	4486	4576	4666	4756	90
4	4845	4935	5025	5114	5204	5294	5383	5473	5563	5652	90
5	5742	5831	5921	6010	6100	6189	6279	6368	6458	6547	89
6	6636	6726	6815	6904	6994	7083	7172	7261	7351	7440	89
7	7529	7618	7707	7796	7886	7975	8064	8153	8242	8331	89
8	8420	8509	8598	8687	8776	8865	8953	9042	9131	9220	89
9	9309	9398	9486	9575	9664	9753	9841	9930	690019	690107	89
490	690196	690285	690373	690462	690550	690639	690727	690816	690905	690993	89
1	1081	1170	1258	1347	1435	1524	1612	1700	1789	1877	88
2	1965	2053	2142	2230	2318	2406	2494	2583	2671	2759	88
3	2847	2935	3023	3111	3199	3287	3375	3463	3551	3639	88
4	3727	3815	3903	3991	4078	4166	4254	4342	4430	4517	88
5	4605	4693	4781	4868	4956	5044	5131	5219	5307	5394	88
6	5482	5569	5657	5744	5832	5919	6007	6094	6182	6269	87
7	6356	6444	6531	6618	6700	6793	6880	6968	7055	7142	87
8	7229	7317	7404	7491	7578	7665	7752	7839	7926	8014	87
9	8101	8188	8275	8362	8449	8535	8622	8709	8796	8883	87
500	698970	699057	699144	699231	699317	699404	699491	699578	699664	699751	87
1	9838	9924	700011	700098	700184	700271	700358	700444	700531	700617	87
2	700704	700790	0877	0963	1050	1136	1222	1309	1395	1482	86
3	1568	1654	1741	1827	1913	1999	2086	2172	2258	2344	86
4	2431	2517	2603	2689	2775	2861	2947	3033	3119	3205	86
5	3291	3377	3463	3549	3635	3721	3807	3893	3979	4065	86
6	4151	4236	4322	4408	4494	4579	4665	4751	4837	4922	86
7	5008	5094	5179	5265	5350	5436	5522	5607	5693	5778	86
8	5864	5949	6035	6120	6206	6291	6376	6462	6547	6632	85
9	6718	6803	6888	6974	7059	7144	7229	7315	7400	7485	85
510	707570	707655	707740	707826	707911	707996	708081	708166	708251	708336	85
1	8421	8506	8591	8676	8761	8846	8931	9015	9100	9185	85
2	9270	9355	9440	9524	9609	9694	9779	9863	9948	710033	85
3	710117	710202	710287	710371	710456	710540	710625	710710	710794	0879	85
4	0963	1048	1132	1217	1301	1385	1470	1554	1639	1723	84
5	1807	1892	1976	2060	2144	2229	2313	2397	2481	2566	84
6	2650	2734	2818	2902	2986	3070	3154	3238	3323	3407	84
7	3491	3575	3659	3742	3826	3910	3994	4078	4162	4246	84
8	4330	4414	4497	4581	4665	4749	4833	4916	5000	5084	84
9	5167	5251	5335	5418	5502	5586	5669	5753	5836	5920	84

| No. | 0 | 1 | 2 | 3 | 4 | 5 | 6 | 7 | 8 | 9 | Diff. |

No.	0	1	2	3	4	5	6	7	8	9	Diff.
520	716003	716087	716170	716254	716337	716421	716504	716588	716671	716754	83
1	6838	6921	7004	7088	7171	7254	7338	7421	7504	7587	83
2	7671	7754	7837	7920	8003	8086	8169	8253	8336	8419	83
3	8502	8585	8668	8751	8834	8917	9000	9083	9165	9248	83
4	9331	9414	9497	9580	9663	9745	9828	9911	9994	720077	83
5	720159	720242	720325	720407	720490	720573	720655	720738	720821	0903	83
6	0986	1068	1151	1233	1316	1398	1481	1563	1646	1728	82
7	1811	1893	1975	2058	2140	2222	2305	2387	2469	2552	82
8	2634	2716	2798	2881	2963	3045	3127	3209	3291	3374	82
9	3456	3538	3620	3702	3784	3866	3948	4030	4112	4194	82
530	724276	724358	724440	724522	724604	724685	724767	724849	724931	725013	82
1	5095	5176	5258	5340	5422	5503	5585	5667	5748	5830	82
2	5912	5993	6075	6156	6238	6320	6401	6483	6564	6646	82
3	6727	6809	6890	6972	7053	7134	7216	7297	7379	7460	81
4	7541	7623	7704	7785	7866	7948	8029	8110	8191	8273	81
5	8354	8435	8516	8597	8678	8759	8841	8922	9003	9084	81
6	9165	9246	9337	9408	9489	9570	9651	9732	9813	9893	81
7	9974	730055	730136	730217	730298	730378	730459	730540	730621	730702	81
8	730782	0863	0944	1024	1105	1186	1266	1347	1428	1508	81
9	1589	1609	1750	1830	1911	1991	2072	2152	2233	2313	81
540	732394	732474	732555	732635	732715	732796	732876	732950	733037	733117	80
1	3197	3278	3358	3438	3518	3598	3679	3759	3839	3919	80
2	3999	4079	4160	4240	4320	4400	4480	4560	4640	4720	80
3	4800	4880	4960	5040	5120	5200	5279	5359	5439	5519	80
4	5599	5679	5759	5838	5918	5998	6078	6157	6237	6317	80
5	6397	6476	6556	6635	6715	6795	6874	6954	7034	7113	80
6	7193	7272	7352	7431	7511	7590	7670	7749	7829	7908	79
7	7987	8067	8146	8225	8305	8384	8463	8543	8622	8701	79
8	8781	8860	8939	9018	9097	9177	9256	9335	9414	9493	79
9	9572	9651	9731	9810	9889	9968	740047	740126	740205	740284	79
550	740363	740442	740521	740600	740678	740757	740836	740915	740994	741073	79
1	1152	1230	1309	1388	1467	1546	1624	1703	1782	1860	79
2	1939	2018	2096	2175	2254	2332	2411	2489	2568	2647	79
3	2725	2804	2882	2961	3039	3118	3196	3275	3353	3431	78
4	3510	3588	3667	3745	3823	3902	3980	4058	4136	4215	78
5	4293	4371	4449	4528	4606	4684	4762	4840	4919	4997	78
6	5075	5153	5231	5309	5387	5465	5543	5621	5699	5777	78
7	5855	5933	6011	6089	6167	6245	6323	6401	6479	6556	78
8	6634	6712	6790	6868	6945	7023	7101	7179	7256	7334	78
9	7412	7489	7567	7645	7722	7800	7878	7955	8033	8110	78
560	748188	748266	748343	748421	748498	748576	748653	748731	748808	748885	77
1	8963	9040	9118	9195	9272	9350	9427	9504	9582	9659	77
2	9736	9814	9891	9968	750045	750123	750200	750277	750354	750431	77
3	750508	750586	750663	750740	0817	0894	0971	1048	1125	1202	77
4	1279	1356	1433	1510	1587	1664	1741	1818	1895	1972	77
5	2048	2125	2202	2279	2356	2433	2509	2586	2663	2740	77
6	2816	2893	2970	3047	3123	3200	3277	3353	3430	3506	77
7	3583	3660	3736	3813	3889	3966	4042	4119	4195	4272	77
8	4348	4425	4501	4578	4654	4730	4807	4883	4960	5036	76
9	5112	5189	5265	5341	5417	5494	5570	5646	5722	5799	76
570	755875	755951	756027	756103	756180	756256	756332	756408	756484	756560	76
1	6636	6712	6788	6864	6940	7016	7092	7168	7244	7320	76
2	7396	7472	7548	7624	7700	7775	7851	7927	8003	8079	76
3	8155	8230	8306	8382	8458	8533	8609	8685	8761	8836	76
4	8912	8988	9063	9139	9214	9290	9366	9441	9517	9592	76
5	9668	9743	9819	9894	9970	760045	760121	760196	760272	760347	75
6	760422	760498	760573	760649	760724	0799	0875	0950	1025	1101	75
7	1176	1251	1326	1402	1477	1552	1627	1702	1778	1853	75
8	1928	2003	2078	2153	2228	2303	2378	2453	2529	2604	75
9	2679	2754	2829	2904	2978	3053	3128	3203	3278	3353	75

No.	0	1	2	3	4	5	6	7	8	9	Diff.

No.	0	1	2	3	4	5	6	7	8	9	Diff.
700	845098	845160	845223	845284	845346	845408	845470	845532	845594	845656	62
1	5718	5780	5842	5904	5966	6028	6090	6151	6213	6275	62
2	6337	6399	6461	6523	6585	6646	6708	6770	6832	6894	62
3	6955	7017	7079	7141	7202	7264	7326	7388	7449	7511	62
4	7573	7634	7696	7758	7819	7881	7943	8004	8066	8128	62
5	8189	8251	8312	8374	8435	8497	8559	8620	8682	8743	62
6	8805	8866	8928	8989	9051	9112	9174	9235	9297	9358	61
7	9419	9481	9542	9604	9065	9726	9788	9849	9911	9972	61
8	850033	850095	850156	850217	850279	850340	850401	850462	850524	850585	61
9	0646	0707	0769	0830	0952	1014	1075	1136	1197		61
710	851258	851319	851381	851442	851503	851564	851625	851686	851747	851809	61
1	1870	1931	1992	2053	2114	2175	2236	2297	2358	2419	61
2	2480	2541	2602	2663	2724	2785	2846	2907	2968	3029	61
3	3090	3150	3211	3272	3333	3394	3455	3516	3577	3637	61
4	3698	3759	3820	3881	3941	4002	4063	4124	4185	4245	61
5	4306	4367	4428	4488	4549	4610	4670	4731	4792	4852	61
6	4913	4974	5034	5095	5156	5216	5277	5337	5398	5459	61
7	5519	5580	5640	5701	5761	5822	5882	5943	6003	6064	61
8	6124	6185	6245	6306	6366	6427	6487	6548	6608	6668	60
9	6729	6789	6850	6910	6970	7031	7091	7152	7212	7272	60
720	857332	857393	857453	857513	857574	857634	857694	857755	857815	857875	60
1	7935	7995	8056	8116	8176	8236	8297	8357	8417	8477	60
2	8537	8597	8657	8718	8778	8838	8898	8958	9018	9078	60
3	9138	9198	9258	9318	9379	9439	9499	9559	9619	9679	60
4	9739	9799	9859	9918	9978	860038	860098	860158	860218	860278	60
5	860338	860398	860458	860518	860578	0637	0697	0757	0817	0877	60
6	0937	0996	1056	1116	1176	1236	1295	1355	1415	1475	60
7	1534	1594	1654	1714	1773	1833	1893	1952	2012	2072	60
8	2131	2191	2251	2310	2370	2430	2489	2549	2608	2668	60
9	2728	2787	2847	2906	2966	3025	3085	3144	3204	3263	60
730	863323	863382	863442	863501	863561	863620	863680	863739	863799	863858	59
1	3917	3977	4036	4096	4155	4214	4274	4333	4392	4452	59
2	4511	4570	4630	4689	4748	4808	4867	4926	4985	5045	59
3	5104	5163	5222	5282	5341	5400	5459	5519	5578	5637	59
4	5696	5755	5814	5874	5933	5992	6051	6110	6169	6228	59
5	6287	6346	6405	6465	6524	6583	6642	6701	6760	6819	59
6	6878	6937	6996	7055	7114	7173	7232	7291	7350	7409	59
7	7467	7526	7585	7644	7703	7762	7821	7880	7939	7998	59
8	8056	8115	8174	8233	8292	8350	8409	8468	8527	8586	59
9	8644	8703	8762	8821	8879	8938	8997	9056	9114	9173	59
740	869232	869290	869349	869408	869466	869525	869584	869642	869701	869760	59
1	9818	9877	9935	9994	870053	870111	870170	870228	870287	870345	59
2	870404	870462	870521	870579	0638	0696	0755	0813	0872	0930	58
3	0989	1047	1106	1164	1223	1281	1339	1398	1456	1515	58
4	1573	1631	1690	1748	1806	1865	1923	1981	2040	2098	58
5	2156	2215	2273	2331	2389	2448	2506	2564	2622	2681	58
6	2739	2797	2855	2913	2972	3030	3088	3146	3204	3262	58
7	3321	3379	3437	3495	3553	3611	3669	3727	3785	3844	58
8	3902	3960	4018	4076	4134	4192	4250	4308	4366	4424	58
9	4482	4540	4598	4656	4714	4772	4830	4888	4945	5003	58
750	875061	875119	875177	875235	875293	875351	875409	875466	875524	875582	58
1	5640	5698	5756	5813	5871	5929	5987	6045	6102	6160	58
2	6218	6276	6333	6391	6449	6507	6564	6622	6680	6737	58
3	6795	6853	6910	6968	7026	7083	7141	7199	7256	7314	58
4	7371	7429	7487	7544	7602	7659	7717	7774	7832	7890	58
5	7947	8004	8062	8119	8177	8234	8292	8349	8407	8464	57
6	8522	8579	8637	8694	8752	8809	8866	8924	8981	9039	57
7	9096	9153	9211	9268	9325	9383	9440	9497	9555	9612	57
8	9669	9726	9784	9841	9898	9956	880013	880070	880127	880185	57
	0.880242	880299	880356	880413	880471	880528	0585	0642	0699	0756	57

| No. | 0 | 1 | 2 | 3 | 4 | 5 | 6 | 7 | 8 | 9 | Diff. |

No.	0	1	2	3	4	5	6	7	8	9	Diff.
760	880814	880871	880928	880985	881042	881099	881156	881213	881271	881328	57
1	1385	1442	1499	1556	1613	1670	1727	1784	1841	1898	57
2	1955	2012	2069	2126	2183	2240	2297	2354	2411	2468	57
3	2525	2581	2638	2695	2752	2809	2866	2923	2980	3037	57
4	3093	3150	3207	3264	3321	3377	3434	3491	3548	3605	57
5	3661	3718	3775	3832	3888	3945	4002	4059	4115	4172	57
6	4229	4285	4342	4399	4455	4512	4569	4625	4682	4739	57
7	4795	4852	4909	4965	5022	5078	5135	5192	5248	5305	57
8	5361	5418	5474	5531	5587	5644	5700	5757	5813	5870	57
9	5926	5983	6039	6096	6152	6209	6265	6321	6378	6434	56
770	886491	886547	886604	886660	886716	886773	886829	886885	886942	886998	56
1	7054	7111	7167	7223	7280	7336	7392	7449	7505	7561	56
2	7617	7674	7730	7786	7842	7898	7955	8011	8067	8123	56
3	8179	8236	8292	8348	8404	8460	8516	8573	8629	8685	56
4	8741	8797	8853	8909	8965	9021	9077	9134	9190	9246	56
5	9302	9358	9414	9470	9526	9582	9638	9694	9750	9806	56
6	9862	9918	9974	890030	890086	890141	890197	890253	890309	890365	56
7	890421	890477	890533	0589	0645	0700	0756	0812	0868	0924	56
8	0980	1035	1091	1147	1203	1259	1314	1370	1426	1482	56
9	1537	1593	1649	1705	1760	1816	1872	1928	1983	2039	56
780	892095	892150	892206	892262	892317	892373	892429	892484	892540	892595	56
1	2651	2707	2762	2818	2873	2929	2985	3040	3096	3151	56
2	3207	3262	3318	3373	3429	3484	3540	3595	3651	3706	56
3	3762	3817	3873	3928	3984	4039	4094	4150	4205	4261	55
4	4316	4371	4427	4482	4538	4593	4648	4704	4759	4814	55
5	4870	4925	4980	5036	5091	5146	5201	5257	5312	5367	55
6	5423	5478	5533	5588	5644	5699	5754	5809	5864	5920	55
7	5975	6030	6085	6140	6195	6251	6306	6361	6416	6471	55
8	6526	6581	6636	6692	6747	6802	6857	6912	6967	7022	55
9	7077	7132	7187	7242	7297	7352	7407	7462	7517	7572	55
790	897627	897682	897737	897792	897847	897902	897957	898012	898067	898122	55
1	8176	8231	8286	8341	8396	8451	8506	8561	8615	8670	55
2	8725	8780	8835	8890	8944	8999	9054	9109	9164	9218	55
3	9273	9328	9383	9437	9492	9547	9602	9656	9711	9766	55
4	9821	9875	9930	9985	900039	900094	900149	900203	900258	900312	55
5	900367	900422	900476	900531	0586	0640	0695	0749	0804	0859	55
6	0913	0968	1022	1077	1131	1186	1240	1295	1349	1404	55
7	1458	1513	1567	1622	1676	1731	1785	1840	1894	1948	54
8	2003	2057	2112	2166	2221	2275	2329	2384	2438	2492	54
9	2547	2601	2655	2710	2764	2818	2873	2927	2981	3036	54
800	903090	903144	903199	903253	903307	903361	903416	903470	903524	903578	54
1	3633	3687	3741	3795	3849	3904	3958	4012	4066	4120	54
2	4174	4229	4283	4337	4391	4445	4499	4553	4607	4661	54
3	4716	4770	4824	4878	4932	4986	5040	5094	5148	5202	54
4	5256	5310	5364	5418	5472	5526	5580	5634	5688	5742	54
5	5796	5850	5904	5958	6012	6066	6119	6173	6227	6281	54
6	6335	6389	6443	6497	6551	6604	6658	6712	6766	6820	54
7	6874	6927	6981	7035	7089	7143	7196	7250	7304	7358	54
8	7411	7465	7519	7573	7626	7680	7734	7787	7841	7895	54
9	7949	8002	8056	8110	8163	8217	8270	8324	8378	8431	54
810	908485	908539	908592	908646	908699	908753	908807	908860	908914	908967	54
1	9021	9074	9128	9181	9235	9289	9342	9396	9449	9503	54
2	9556	9610	9663	9716	9770	9823	9877	9930	9984	910037	53
3	910091	910144	910197	910251	910304	910358	910411	910464	910518	0571	53
4	0624	0678	0731	0784	0838	0891	0944	0998	1051	1104	53
5	1158	1211	1264	1317	1371	1424	1477	1530	1584	1637	53
6	1690	1743	1797	1850	1903	1956	2009	2063	2116	2169	53
7	2222	2275	2328	2381	2435	2488	2541	2594	2647	2700	53
8	2753	2806	2859	2913	2966	3019	3072	3125	3178	3231	53
9	3284	3337	3390	3443	3496	3549	3602	3655	3708	3761	53
No.	0	1	2	3	4	5	6	7	8	9	Diff.

No.	0	1	2	3	4	5	6	7	8	9	Diff.
820	913814	913867	913920	913973	914026	914079	914132	914184	914237	914290	53
1	4343	4396	4449	4502	4555	4608	4660	4713	4768	4819	53
2	4872	4925	4977	5030	5083	5136	5189	5241	5294	5347	53
3	5400	5453	5505	5558	5611	5664	5716	5769	5822	5875	53
4	5927	5980	6033	6085	6138	6191	6243	6296	6349	6401	53
5	6454	6507	6559	6612	6664	6717	6770	6822	6875	6927	53
6	6980	7033	7085	7138	7190	7243	7295	7348	7400	7453	53
7	7506	7558	7611	7663	7716	7768	7820	7873	7925	7978	52
8	8030	8083	8135	8188	8240	8293	8345	8397	8450	8502	52
9	8555	8607	8659	8712	8764	8816	8869	8921	8973	9026	52
830	919078	919130	919183	919235	919287	919340	919392	919444	919496	919549	52
1	9601	9653	9706	9758	9810	9862	9914	9967	920019	920071	52
2	920123	920176	920228	920280	920332	920384	920436	920489	0541	0593	52
3	0645	0697	0749	0801	0853	0906	0958	1010	1062	1114	52
4	1166	1218	1270	1322	1374	1426	1478	1530	1582	1634	52
5	1686	1738	1790	1842	1894	1946	1998	2050	2102	2154	52
6	2206	2258	2310	2362	2414	2466	2518	2570	2622	2674	52
7	2725	2777	2829	2881	2933	2985	3037	3089	3140	3192	52
8	3244	3296	3348	3399	3451	3503	3555	3607	3658	3710	52
9	3762	3814	3865	3917	3969	4021	4072	4124	4176	4228	52
840	924279	924331	924383	924434	924486	924538	924589	924641	924693	924744	52
1	4796	4848	4899	4951	5003	5054	5106	5157	5209	5261	52
2	5312	5364	5415	5467	5518	5570	5621	5673	5725	5776	52
3	5828	5879	5931	5982	6034	6085	6137	6188	6240	6291	51
4	6342	6394	6445	6497	6548	6600	6651	6702	6754	6805	51
5	6857	6908	6959	7011	7062	7114	7165	7216	7268	7319	51
6	7370	7422	7473	7524	7576	7627	7678	7730	7781	7832	51
7	7883	7935	7986	8037	8088	8140	8191	8242	8293	8345	51
8	8396	8447	8498	8549	8601	8652	8703	8754	8805	8857	51
9	8908	8959	9010	9061	9112	9163	9215	9266	9317	9368	51
850	929419	929470	929521	929572	929623	929674	929725	929776	929827	929879	51
1	9930	9981	930032	930083	930134	930185	930236	930287	930338	930389	51
2	930440	930491	0542	0592	0643	0694	0745	0796	0847	0898	51
3	0949	1000	1051	1102	1153	1204	1254	1305	1356	1407	51
4	1458	1509	1560	1610	1661	1712	1763	1814	1865	1915	51
5	1966	2017	2068	2118	2169	2220	2271	2322	2372	2423	51
6	2474	2524	2575	2626	2677	2727	2778	2829	2879	2930	51
7	2981	3031	3082	3133	3183	3234	3285	3335	3386	3437	51
8	3487	3538	3589	3639	3690	3740	3791	3841	3892	3943	51
9	3993	4044	4094	4145	4195	4246	4296	4347	4397	4448	51
860	934498	934549	934599	934650	934700	934751	934801	934852	934902	934953	50
1	5003	5054	5104	5154	5205	5255	5306	5356	5406	5457	50
2	5507	5558	5608	5658	5709	5759	5809	5860	5910	5960	50
3	6011	6061	6111	6162	6212	6262	6313	6363	6413	6463	50
4	6514	6564	6614	6665	6715	6765	6815	6865	6916	6966	50
5	7016	7066	7117	7167	7217	7267	7317	7367	7418	7468	50
6	7518	7568	7618	7668	7718	7769	7819	7869	7919	7969	50
7	8019	8069	8119	8169	8219	8269	8320	8370	8420	8470	50
8	8520	8570	8620	8670	8720	8770	8820	8870	8920	8970	50
9	9020	9070	9120	9170	9220	9270	9320	9369	9419	9469	50
870	939519	939569	939619	939669	939719	939769	939819	939869	939918	939968	50
1	940018	940068	940118	940168	940218	940267	940317	940367	940417	940467	50
2	0516	0566	0616	0666	0716	0765	0815	0865	0915	0964	50
3	1014	1064	1114	1163	1213	1263	1313	1362	1412	1462	50
4	1511	1561	1611	1660	1710	1760	1809	1859	1909	1958	50
5	2008	2058	2107	2157	2207	2256	2306	2355	2405	2455	50
6	2504	2554	2603	2653	2702	2752	2801	2851	2901	2950	50
7	3000	3049	3099	3148	3198	3247	3297	3346	3396	3445	49
8	3495	3544	3593	3643	3692	3742	3791	3841	3890	3939	49
9	3989	4038	4088	4137	4186	4236	4285	4335	4384	4433	49
No.	0	1	2	3	4	5	6	7	8	9	Diff.

No.	0	1	2	3	4	5	6	7	8	9	Dif.
880	944483	944532	944581	944631	944680	944729	944779	944828	944877	944927	49
1	4976	5025	5074	5124	5173	5222	5272	5321	5370	5419	49
2	5469	5518	5567	5616	5665	5715	5764	5813	5862	5912	49
3	5961	6010	6059	6108	6157	6207	6256	6305	6354	6403	49
4	6452	6501	6551	6600	6649	6698	6747	6796	6845	6894	49
5	6943	6992	7041	7090	7140	7189	7238	7287	7336	7385	49
6	7434	7483	7532	7581	7630	7679	7728	7777	7826	7875	49
7	7924	7973	8022	8070	8119	8168	8217	8266	8315	8364	49
8	8413	8462	8511	8560	8609	8657	8706	8755	8804	8853	49
9	8902	8951	8999	9048	9097	9146	9195	9244	9292	9341	49
890	949390	949439	949488	949536	949585	949634	949683	949731	949780	949829	49
1	9878	9926	9975	950024	950073	950121	950170	950219	950267	950316	49
2	950365	950414	950462	0511	0560	0608	0657	0706	0754	0803	49
3	0851	0900	0949	0997	1046	1095	1143	1192	1240	1289	49
4	1338	1386	1435	1483	1532	1580	1629	1677	1726	1775	49
5	1823	1872	1920	1969	2017	2066	2114	2163	2211	2260	49
6	2308	2356	2405	2453	2502	2550	2599	2647	2696	2744	48
7	2702	2841	2889	2938	2986	3034	3083	3131	3180	3228	48
8	3276	3325	3373	3421	3470	3518	3566	3615	3663	3711	48
9	3760	3808	3856	3905	3953	4001	4049	4098	4146	4194	48
900	954243	954291	954339	954387	954435	954484	954532	954580	954628	954676	48
1	4725	4773	4821	4869	4918	4966	5014	5062	5110	5158	48
2	5207	5255	5303	5351	5399	5447	5495	5543	5592	5640	48
3	5688	5736	5784	5832	5880	5928	5976	6024	6072	6120	48
4	6168	6216	6265	6313	6361	6409	6457	6505	6553	6601	48
5	6649	6697	6745	6793	6840	6888	6936	6984	7032	7080	48
6	7128	7176	7224	7272	7320	7368	7416	7464	7512	7559	48
7	7607	7655	7703	7751	7799	7847	7894	7942	7990	8038	48
8	8086	8134	8181	8229	8277	8325	8373	8421	8468	8516	48
9	8564	8612	8659	8707	8755	8803	8850	8898	8946	8994	48
910	959041	959089	959137	959185	959232	959280	959328	959375	959423	959471	48
1	9518	9566	9614	9661	9709	9757	9804	9852	9900	9947	48
2	9995	960042	960090	960138	960185	960233	960281	960328	960376	960423	48
3	960471	0518	0566	0613	0661	0709	0756	0804	0851	0899	48
4	0946	0994	1041	1089	1136	1184	1231	1279	1326	1374	48
5	1421	1469	1516	1563	1611	1658	1706	1753	1801	1848	47
6	1895	1943	1990	2038	2085	2132	2180	2227	2275	2322	47
7	2369	2417	2464	2511	2559	2606	2653	2701	2748	2795	47
8	2843	2890	2937	2985	3032	3079	3126	3174	3221	3268	47
9	3316	3363	3410	3457	3504	3552	3599	3646	3693	3741	47
920	963788	963835	963882	963929	963977	964024	964071	964118	964165	964212	47
1	4260	4307	4354	4401	4448	4495	4542	4590	4637	4684	47
2	4731	4778	4825	4872	4919	4966	5013	5061	5108	5155	47
3	5202	5249	5296	5343	5390	5437	5484	5531	5578	5625	47
4	5672	5719	5766	5813	5860	5907	5954	6001	6048	6095	47
5	6142	6189	6236	6283	6329	6376	6423	6470	6517	6564	47
6	6611	6658	6705	6752	6799	6845	6892	6939	6986	7033	47
7	7080	7127	7173	7220	7267	7314	7361	7408	7454	7501	47
8	7548	7595	7642	7688	7735	7782	7829	7875	7922	7969	47
9	8016	8062	8109	8156	8203	8249	8296	8343	8390	8436	47
930	968483	968530	968576	968623	968670	968716	968763	968810	968856	968903	47
1	8950	8996	9043	9090	9136	9183	9229	9276	9323	9369	47
2	9416	9463	9509	9556	9602	9649	9695	9742	9789	9835	47
3	9882	9928	9975	970021	970068	970114	970161	970207	970254	970300	47
4	970347	970393	970440	0486	0533	0579	0626	0672	0719	0765	46
5	0812	0858	0904	0951	0997	1044	1090	1137	1183	1229	46
6	1276	1322	1369	1415	1461	1508	1554	1601	1647	1693	46
7	1740	1786	1832	1879	1925	1971	2018	2064	2110	2157	46
8	2203	2249	2295	2342	2388	2434	2481	2527	2573	2619	46
9	2666	2712	2758	2804	2851	2897	2943	2989	3035	3082	46
No.	0	1	2	3	4	5	6	7	8	9	Dif.

No.	0	1	2	3	4	5	6	7	8	9	Dff.
940	973128	973174	973220	973266	973313	973359	973405	973451	973497	973543	46
1	3590	3636	3682	3728	3774	3820	3866	3913	3959	4005	46
2	4051	4097	4143	4189	4235	4281	4327	4374	4420	4466	46
3	4512	4558	4604	4650	4696	4742	4788	4834	4880	4926	46
4	4972	5018	5064	5110	5156	5202	5248	5294	5340	5386	46
5	5432	5478	5524	5570	5616	5662	5707	5753	5799	5845	46
6	5891	5937	5983	6029	6075	6121	6167	6212	6258	6304	46
7	6350	6396	6442	6488	6533	6570	6625	6671	6717	6763	46
8	6808	6854	6900	6946	6992	7037	7083	7129	7175	7220	46
9	7260	7312	7358	7403	7449	7495	7541	7586	7632	7678	46
950	977724	977769	977815	977861	977906	977952	977998	978043	978089	978135	46
1	8181	8226	8272	8317	8363	8409	8454	8500	8546	8591	46
2	8637	8683	8728	8774	8819	8865	8911	8956	9002	9047	46
3	9093	9138	9184	9230	9275	9321	9366	9412	9457	9503	46
4	9548	9594	9639	9685	9730	9776	9821	9867	9912	9958	46
5	980003	980049	980094	980140	980185	980231	980276	980322	980367	980412	45
6	0458	0503	0549	0594	0640	0685	0730	0776	0821	0867	45
7	0912	0957	1003	1048	1093	1139	1184	1229	1275	1320	45
8	1366	1411	1456	1501	1547	1592	1637	1683	1728	1773	45
9	1819	1864	1909	1954	2000	2045	2090	2135	2181	2226	45
960	982271	982316	982362	982407	982452	982497	982543	982588	982633	982678	45
1	2723	2769	2814	2859	2904	2949	2994	3040	3085	3130	45
2	3175	3220	3265	3310	3356	3401	3446	3491	3536	3581	45
3	3626	3671	3716	3762	3807	3852	3897	3942	3987	4032	45
4	4077	4122	4167	4212	4257	4302	4347	4392	4437	4482	45
5	4527	4572	4617	4662	4707	4752	4797	4842	4887	4932	45
6	4977	5022	5067	5112	5157	5202	5247	5292	5337	5382	45
7	5426	5471	5516	5561	5606	5651	5696	5741	5786	5830	45
8	5875	5920	5965	6010	6055	6100	6144	6189	6234	6279	45
9	6324	6369	6413	6458	6503	6548	6593	6637	6682	6727	45
970	986772	986817	986861	986906	986951	986996	987040	987085	987130	987175	45
1	7219	7264	7309	7353	7398	7443	7488	7532	7577	7622	45
2	7666	7711	7756	7800	7845	7890	7934	7979	8024	8068	45
3	8113	8157	8202	8247	8291	8336	8381	8425	8470	8514	45
4	8559	8604	8648	8693	8737	8782	8826	8871	8916	8960	45
5	9005	9049	9094	9138	9183	9227	9272	9316	9361	9405	45
6	9450	9494	9539	9583	9628	9672	9717	9761	9806	9850	44
7	9895	9939	9983	990028	990072	990117	990161	990206	990250	990294	44
8	990339	990383	990428	0472	0516	0561	0605	0650	0694	0738	44
9	0783	0827	0871	0916	0960	1004	1049	1093	1137	1182	44
980	991226	991270	991315	991359	991403	991448	991492	991536	991580	991625	44
1	1669	1713	1758	1802	1846	1890	1935	1979	2023	2067	44
2	2111	2156	2200	2244	2288	2333	2377	2421	2465	2509	44
3	2554	2598	2642	2686	2730	2774	2819	2863	2907	2951	44
4	2995	3039	3083	3127	3172	3216	3260	3304	3348	3392	44
5	3436	3480	3524	3568	3613	3657	3701	3745	3789	3833	44
6	3877	3921	3965	4009	4053	4097	4141	4185	4229	4273	44
7	4317	4361	4405	4449	4493	4537	4581	4625	4669	4713	44
8	4757	4801	4845	4889	4933	4977	5021	5065	5108	5152	44
9	5196	5240	5284	5328	5372	5416	5460	5504	5547	5591	44
990	995635	995679	995723	995767	995811	995854	995898	995942	995986	996030	44
1	6074	6117	6161	6205	6249	6293	6337	6380	6424	6468	44
2	6512	6555	6599	6643	6687	6731	6774	6818	6862	6906	44
3	6949	6993	7037	7080	7124	7168	7212	7255	7299	7343	44
4	7386	7430	7474	7517	7561	7605	7648	7692	7736	7779	44
5	7823	7867	7910	7954	7998	8041	8085	8129	8172	8216	44
6	8259	8303	8347	8390	8434	8477	8521	8564	8608	8652	44
7	8695	8739	8782	8826	8869	8913	8956	9000	9043	9087	44
8	9131	9174	9218	9261	9305	9348	9392	9435	9479	9522	44
9	9565	9600	9652	9696	9739	9783	9826	9870	9913	9957	43
No.	0	1	2	3	4	5	6	7	8	9	Dff.

LOGARITHMIC

SINES AND TANGENTS,

FOR EVERY

DEGREE AND MINUTE

OF

THE QUADRANT.

N. B. The minutes in the left-hand column of each page, increasing downwards, belong to the degrees at the top; and those increasing upwards, in the right-hand column, belong to the degrees below.

M.	Sine	D.	Cosine	D.	Tang.	D.	Cotang.	
0	0·000000		10·000000		0·000000		Infinite.	60
1	6·463726	501717	000000	00	0·463726	501717	13·536274	59
2	764756	293485	000000	00	764756	293483	235244	58
3	940847	208231	000000	00	940847	208231	059153	57
4	7·065786	161517	000000	00	7·065786	161517	12·934214	56
5	162696	131968	000000	00	162696	131969	837304	55
6	241877	111575	9·999999	01	241878	111578	758122	54
7	308894	96653	999999	01	308895	99653	691175	53
8	366816	85254	999999	01	366817	85254	633183	52
9	417968	76263	999999	01	417970	76263	582030	51
10	463725	68088	999998	01	463727	68088	536273	50
11	7·505118	62981	9·999998	01	7·505120	62981	12·494880	49
12	542906	57936	999997	01	542909	57933	457091	48
13	577668	53641	999997	01	577672	53642	422328	47
14	609853	49938	999996	01	609857	49939	390143	46
15	639816	46714	999996	01	639820	46715	360180	45
16	667845	43881	999995	01	667849	43882	332151	44
17	694173	41372	999995	01	694179	41373	305821	43
18	718997	39135	999994	01	719003	39136	280997	42
19	742477	37127	999993	01	742484	37128	257516	41
20	764754	35315	999993	01	764761	35316	235239	40
21	7·785943	33672	9·999992	01	7·785951	33673	12·214049	39
22	806146	32175	999991	01	806155	32176	193845	38
23	825451	30805	999990	01	825460	30806	174540	37
24	843934	29547	999989	02	843944	29549	156056	36
25	861662	28388	999988	02	861674	28390	138326	35
26	878695	27317	999988	02	878708	27318	121292	34
27	895085	26323	999987	02	895099	26325	104901	33
28	910879	25399	999986	02	910894	25401	089106	32
29	926119	24538	999985	02	926134	24540	073860	31
30	940842	23733	999983	02	940858	23735	059142	30
31	7·955082	22980	9·999982	02	7·955100	22981	12·044900	29
32	968870	22273	999981	02	968889	22275	031111	28
33	982233	21608	999980	02	982253	21610	017747	27
34	995198	20981	999979	02	995219	20983	004781	26
35	8·007787	20390	999977	02	8·007809	20392	11·992191	25
36	020021	19831	999976	02	020045	19833	979955	24
37	031919	19302	999975	02	031945	19305	968055	23
38	043501	18801	999973	02	043527	18803	956473	22
39	054781	18325	999972	02	054809	18327	945191	21
40	065776	17872	999971	02	065806	17874	934194	20
41	8·076500	17441	9·999969	02	8·076531	17444	11·923469	19
42	086965	17031	999968	02	086997	17034	913003	18
43	097183	16639	999966	02	007217	16642	902783	17
44	107167	16265	999964	03	107202	16268	892797	16
45	116926	15908	999963	03	116963	15910	883037	15
46	126471	15566	999961	03	126510	15568	873490	14
47	135810	15238	999959	03	135851	15241	864149	13
48	144953	14924	999958	03	144996	14927	855004	12
49	153907	14622	999956	03	153952	14627	846048	11
50	162681	14333	999954	03	162727	14336	837273	10
51	8·171280	14054	9·999952	03	8·171328	14057	11·828672	9
52	179713	13786	999950	03	179763	13700	820237	8
53	187985	13529	999948	03	188036	13532	811964	7
54	196102	13280	999946	03	196156	13284	803844	6
55	204070	13041	999944	03	204126	13044	795874	5
56	211895	12810	999942	04	211953	12814	788047	4
57	219581	12587	999940	04	219641	12590	780359	3
58	227134	12372	999938	04	227195	12376	772805	2
59	234557	12164	999936	04	234621	12168	765379	1
60	241855	11963	999934	04	241921	11967	758079	0
	Cosine		Sine		Cotang.		Tang.	

89 Degrees.

M.	Sine	D.	Cosine	D.	Tang.	D.	Cotang.	
0	8·241855	11963	9·999934	04	8·241921	11967	11·758079	60
1	,249033	11768	999932	04	249102	11772	750898	59
2	256094	11580	999929	04	256165	11584	743835	58
3	263042	11398	999927	04	263115	11402	736885	57
4	269881	11221	999925	04	269956	11225	730044	56
5	276614	11050	999922	04	276691	11054	723309	55
6	283243	10883	999920	04	283323	10887	716677	54
7	289773	10721	999918	04	289856	10726	710144	53
8	296207	10565	999915	04	296292	10570	703708	52
9	302546	10413	999913	04	302634	10418	697366	51
10	308794	10266	999910	04	308884	10270	691116	50
11	8·314954	10122	9·999907	04	8·315046	10126	11·684954	49
12	321027	9982	999905	04	321122	9987	678878	48
13	327016	9847	999902	04	327114	9851	672886	47
14	332924	9714	999899	05	333025	9719	666975	46
15	338753	9586	999897	05	338856	9590	661144	45
16	344504	9460	999894	05	344610	9465	655390	44
17	350181	9338	999891	05	350289	9343	649711	43
18	355783	9219	999888	05	355895	9224	644105	42
19	361315	9103	999885	05	361430	9108	638570	41
20	366777	8990	999882	05	366895	8995	633105	40
21	8·372171	8880	9·999879	05	8·372292	8885	11·627708	39
22	377499	8772	999876	05	377622	8777	622378	38
23	382762	8667	999873	05	382889	8672	617111	37
24	387962	8564	999870	05	388092	8570	611908	36
25	393101	8464	999867	05	393234	8470	606766	35
26	398179	8366	999864	05	398315	8371	601685	34
27	403199	8271	999861	05	403338	8276	596662	33
28	408161	8177	999858	05	408304	8182	591696	32
29	413068	8086	999854	05	413213	8091	586787	31
30	417919	7996	999851	06	418068	8002	581932	30
31	8·422717	7909	9·999848	06	8·422869	7914	11·577131	29
32	427462	7823	999844	06	427618	7830	572382	28
33	432156	7740	999841	06	432315	7745	567685	27
34	436800	7657	999838	06	436962	7663	563038	26
35	441394	7577	999834	06	441560	7583	558440	25
36	445941	7499	999831	06	446110	7505	553890	24
37	450440	7422	999827	06	450613	7428	549387	23
38	454893	7346	999823	06	455070	7352	544930	22
39	459301	7273	999820	06	459481	7279	540519	21
40	463665	7200	999816	06	463849	7206	536151	20
41	8·467985	7129	9·999812	06	8·468172	7135	11·531828	19
42	472263	7060	999809	06	472454	7066	527546	18
43	476498	6991	999805	06	476693	6998	523307	17
44	480693	6924	999801	06	480892	6931	519108	16
45	484848	6859	999797	07	485050	6865	514950	15
46	488963	6794	999793	07	489170	6801	510830	14
47	493040	6731	999790	07	493250	6738	506750	13
48	497078	6669	999786	07	497293	6676	502707	12
49	501080	6608	999782	07	501298	6615	498702	11
50	505045	6548	999778	07	505267	6555	494733	10
51	8·508974	6489	9·999774	07	8·509200	6496	11·490800	9
52	512867	6431	999769	07	513098	6439	486902	8
53	516726	6375	999765	07	516961	6382	483039	7
54	520551	6319	999761	07	520790	6326	479210	6
55	524343	6264	999757	07	524586	6272	475414	5
56	528102	6211	999753	07	528349	6218	471651	4
57	531828	6158	999748	07	532080	6165	467920	3
58	535523	6106	999744	07	535779	6113	464221	2
59	539188	6055	999740	07	539447	6062	460553	1
60	542819	6004	999735	07	543084	6012	456916	0
	Cosine		Sine		Cotang.		Tang.	M.

88 Degrees.

M.	Sine	D.	Cosine	D.	Tang.	D.	Cotang.	
0	8·542819	6004	9·999735	07	8·543084	6012	11·456916	60
1	546422	5955	999731	07	546691	5963	453309	59
2	549995	5906	999728	07	550268	5914	449732	58
3	553539	5858	999722	08	553817	5866	446183	57
4	557054	5811	999717	08	557330	5819	442004	56
5	560540	5765	999713	08	560628	5773	439172	55
6	563999	5719	999708	08	564291	5727	435709	54
7	567431	5674	999704	08	567727	5682	432273	53
8	570836	5630	999699	08	571137	5638	428863	52
9	574214	5587	999694	08	574520	5595	425480	51
10	577566	5544	999689	08	577877	5552	422123	50
11	8·580892	5502	9·999685	08	8·581208	5510	11·418792	49
12	584193	5460	999680	08	584514	5468	415486	48
13	587469	5419	999675	08	587795	5427	412205	47
14	590721	5379	999670	08	591051	5387	408949	46
15	593948	5339	999665	08	594283	5347	405717	45
16	597152	5300	999660	08	597492	5308	402508	44
17	600332	5261	999655	08	600677	5270	399323	43
18	603489	5223	999650	08	603839	5232	396161	42
19	606623	5186	999645	09	606978	5194	393022	41
20	609734	5149	999640	09	610094	5158	389906	40
21	8·612823	5112	9·999635	09	8·613189	5121	11·386811	39
22	615891	5076	999629	09	616262	5085	383738	38
23	618937	5041	999624	09	619313	5050	380687	37
24	621962	5006	999619	09	622343	5015	377657	36
25	624965	4972	999614	09	625352	4981	374648	35
26	627948	4938	999608	09	628340	4947	371660	34
27	630911	4904	999603	09	631308	4913	368692	33
28	633854	4871	999597	09	634256	4880	365744	32
29	636776	4839	999592	09	637184	4848	362816	31
30	639680	4806	999586	09	640093	4816	359907	30
31	8·642563	4775	9·999581	09	8·642982	4784	11·357018	29
32	645428	4743	999575	09	645853	4753	354147	28
33	648274	4712	999570	09	648704	4722	351296	27
34	651102	4682	999564	09	651537	4691	348463	26
35	653911	4652	999558	10	654352	4661	345648	25
36	656702	4622	999553	10	657149	4631	342851	24
37	659475	4592	999547	10	659928	4602	340072	23
38	662230	4563	999541	10	662689	4573	337311	22
39	664968	4535	999535	10	665433	4544	334567	21
40	667689	4506	999529	10	668160	4526	331840	20
41	8·670393	4479	9·999524	10	8·670870	4489	11·329130	19
42	673080	4451	999518	10	673563	4461	326437	18
43	675751	4424	999512	10	676239	4434	323761	17
44	678405	4397	999506	10	678900	4417	321100	16
45	681043	4370	999500	10	681544	4380	318456	15
46	683665	4344	999493	10	684172	4354	315828	14
47	686272	4318	999487	10	686784	4328	313216	13
48	688863	4292	999481	10	689381	4303	310619	12
49	691438	4267	999475	10	691963	4277	308037	11
50	693998	4242	999469	10	694529	4252	305471	10
51	8·696543	4217	9·999463	11	8·697081	4228	11·302919	9
52	699073	4192	999456	11	699617	4203	300383	8
53	701589	4168	999450	11	702139	4179	297861	7
54	704090	4144	999443	11	704646	4155	295354	6
55	706577	4121	999437	11	707140	4132	292860	5
56	709049	4097	999431	11	709618	4108	290382	4
57	711507	4074	999424	11	712083	4085	287917	3
58	713952	4051	999418	11	714534	4062	285465	2
59	716383	4029	999411	11	716972	4040	283028	1
60	718800	4006	999404	11	719396	4017	280604	0
	Cosine		Sine		Cotang.		Tang.	M.

87 Degrees.

M.	Sine	D.	Cosine	D.	Tang.	D.	Cotang.	
0	8·718800	4008	9·999404	11	8·719396	4017	11·280604	60
1	721204	3984	999398	11	721806	3995	278194	59
2	723595	3962	999391	11	724204	3974	275796	58
3	725972	3941	999384	11	726588	3952	273412	57
4	728337	3919	999378	11	728959	3930	271041	56
5	730688	3898	999371	11	731317	3909	268683	55
6	733027	3877	999364	12	733663	3889	266337	54
7	735354	3857	999357	12	735996	3868	264004	53
8	737667	3836	999350	12	738317	3848	261683	52
9	739969	3816	999343	12	740626	3827	259374	51
10	742259	3796	999336	12	742922	3807	257078	50
11	8·744536	3776	9·999329	12	8·745207	3787	11·254793	49
12	746802	3756	999322	12	747479	3768	252521	48
13	749055	3737	999315	12	749740	3749	250260	47
14	751297	3717	999308	12	751989	3729	248011	46
15	753528	3698	999301	12	754227	3710	245773	45
16	755747	3679	999294	12	756453	3692	243547	44
17	757955	3661	999286	12	758668	3673	241332	43
18	760151	3642	999279	12	760872	3655	239128	42
19	762337	3624	999272	12	763065	3636	236935	41
20	704511	3606	999265	12	765246	3618	234754	40
21	8·766675	3588	9·999257	12	8·767417	3600	11·232583	39
22	768828	3570	999250	13	769578	3583	230422	38
23	770970	3553	999242	13	771727	3565	228273	37
24	773101	3535	999235	13	773866	3548	226134	36
25	775223	3518	999227	13	775995	3531	224005	35
26	777333	3501	999220	13	778114	3514	221886	34
27	779434	3484	999212	13	780222	3497	219778	33
28	781524	3467	999205	13	782320	3480	217680	32
29	783605	3451	999197	13	784408	3464	215592	31
30	785675	3431	999189	13	786486	3447	213514	30
31	8·787736	3418	9·999181	13	8·788554	3431	11·211446	29
32	789787	3402	999174	13	790613	3414	209387	28
33	791828	3386	999166	13	792662	3399	207338	27
34	793859	3370	999158	13	794701	3383	205299	26
35	795881	3354	999150	13	796731	3368	203269	25
36	797894	3339	999142	13	798752	3352	201248	24
37	799897	3323	999134	13	800763	3337	199237	23
38	801892	3308	999126	13	802765	3322	197235	22
39	803876	3293	999118	13	804758	3307	195242	21
40	805852	3278	999110	13	806742	3292	193258	20
41	8·807819	3263	9·999102	13	8·808717	3278	11·191283	19
42	809777	3249	999094	14	810683	3262	189317	18
43	811726	3234	999086	14	812641	3248	187359	17
44	813667	3219	999077	14	814589	3233	185411	16
45	815599	3205	999069	14	816529	3219	183471	15
46	817522	3191	999061	14	818461	3205	181539	14
47	819436	3177	999053	14	820384	3191	179616	13
48	821343	3163	999044	14	822298	3177	177702	12
49	823240	3149	999036	14	824205	3163	175795	11
50	825130	3135	999027	14	826103	3150	173897	10
51	8·827011	3122	9·999019	14	8·827992	3136	11·172008	9
52	828884	3108	999010	14	829874	3123	170126	8
53	830749	3095	999002	14	831748	3110	168252	7
54	832607	3082	998993	14	833613	3096	166387	6
55	834456	3069	998984	14	835471	3083	164529	5
56	836297	3056	998976	14	837321	3070	162679	4
57	838130	3043	998967	15	839163	3057	160837	3
58	839956	3030	998958	15	840998	3045	159002	2
59	841774	3017	998950	15	842825	3032	157175	1
60	843585	3000	998941	15	844644	3019	155356	0
	Cosine		Sine		Cotang.		Tang.	M.

86 Degrees.

M.	Sine	D.	Cosine	D.	Tang.	D.	Cotang.	
0	8·843585	3005	9·998941	15	8·844644	3019	11·155356	60
1	845387	2992	998932	15	846455	3007	153545	59
2	847183	2980	998923	15	848260	2995	151740	58
3	848971	2967	998914	15	850057	2982	149943	57
4	850751	2955	998905	15	851846	2970	148154	56
5	852523	2943	998896	15	853628	2958	146372	55
6	854291	2931	998887	15	855403	2946	144597	54
7	856049	2919	998878	15	857171	2935	142829	53
8	857801	2907	998869	15	858932	2923	141068	52
9	859546	2896	998860	15	860686	2911	139314	51
10	861283	2884	998851	15	862433	2900	137567	50
11	8·863014	2873	9·998841	15	8·864173	2888	11·135827	49
12	864738	2861	998832	15	865906	2877	134094	48
13	866455	2850	998823	16	867632	2866	132368	47
14	868165	2839	998813	16	869351	2854	130649	46
15	869868	2828	998804	16	871064	2843	128936	45
16	871565	2817	998795	16	872770	2832	127230	44
17	873255	2806	998785	16	874469	2821	125531	43
18	874938	2795	998770	16	876162	2811	123838	42
19	876615	2786	998766	16	877849	2800	122151	41
20	878285	2773	998757	16	879529	2789	120471	40
21	8·879949	2763	9·998747	16	8·881209	2779	11·118798	39
22	881607	2752	998738	16	882969	2768	117131	38
23	883258	2742	998728	16	884530	2758	115470	37
24	884903	2731	998718	16	886185	2747	113815	36
25	886542	2721	998708	16	887833	2737	112167	35
26	888174	2711	998699	16	889476	2727	110524	34
27	889801	2700	998689	16	891112	2717	108888	33
28	891421	2690	998679	16	892742	2707	107258	32
29	893035	2680	998669	17	894366	2697	105634	31
30	894643	2670	998659	17	895984	2687	104016	30
31	8·896246	2660	9·998649	17	8·897596	2677	11·102404	29
32	897842	2651	998639	17	899203	2667	100797	28
33	899432	2641	998629	17	900803	2658	099197	27
34	901017	2631	998619	17	902398	2648	097602	26
35	902596	2622	998609	17	903987	2638	096013	25
36	904169	2612	998590	17	905570	2629	094430	24
37	905736	2603	998580	17	907147	2620	092853	23
38	907297	2593	998578	17	908719	2610	091281	22
39	908853	2584	998568	17	910285	2601	089715	21
40	910404	2575	998558	17	911846	2592	088154	20
41	8·911049	2566	9·998548	17	8·913401	2583	11·096509	19
42	913488	2556	998537	17	914951	2574	085049	18
43	915022	2547	998527	17	916495	2565	083505	17
44	916550	2538	998516	18	918034	2556	081966	16
45	918073	2529	998506	18	919568	2547	080432	15
46	919591	2520	998495	18	921096	2538	078904	14
47	921103	2512	998485	18	922619	2530	077381	13
48	922610	2503	998474	18	924136	2521	075864	12
49	924112	2494	998464	18	925649	2512	074351	11
50	925609	2486	998453	18	927156	2503	072844	10
51	8·927100	2477	9·998442	18	8·928658	2495	11·071342	9
52	928587	2460	998431	18	930155	2486	069845	8
53	930068	2460	998421	18	931647	2478	068353	7
54	931544	2452	998410	18	933134	2470	066866	6
55	933015	2443	998399	18	934616	2461	065384	5
56	934481	2435	998388	18	936093	2453	063907	4
57	935942	2427	998377	18	937565	2445	062435	3
58	937398	2419	998366	18	939032	2437	060968	2
59	938850	2411	998355	18	940494	2430	059506	1
60	940296	2403	998344	18	941952	2421	058048	0
	Cosine		Sine		Cotang.		Tang.	M.

M.	Sine	D.	Cosine	D.	Tang.	D.	Cotang.	
0	8·940296	2403	9·998344	19	8·941952	2421	11·058048	60
1	941738	2394	998333	19	943404	2413	056596	59
2	943174	2387	998323	19	944852	2405	055148	58
3	944606	2379	998311	19	946295	2397	053705	57
4	946034	2371	998300	19	947734	2390	052266	56
5	947456	2363	998289	19	949168	3282	050832	55
6	948874	2355	998277	19	950597	2374	049403	54
7	950287	2348	998266	19	952021	2366	047979	53
8	951696	2340	998255	19	953441	2360	046559	52
9	953100	2332	998243	19	954856	2351	045144	51
10	954499	2325	998232	19	956267	2344	043733	50
11	8·955894	2317	9·998220	19	8·957674	2337	11·042326	49
12	957284	2310	998209	19	959075	2329	040925	48
13	958670	2302	998197	19	960473	2323	039527	47
14	960052	2295	998186	19	961866	2314	038134	46
15	961429	2288	998174	19	963255	2307	036745	45
16	962801	2280	998163	19	964639	2300	035361	44
17	964170	2273	998151	19	966010	2293	033981	43
18	965534	2266	998139	20	967394	2286	032606	42
19	966893	2259	998128	20	968766	2279	031234	41
20	968249	2252	998116	20	970133	2271	029867	40
21	8·969600	2244	9·998104	20	8·971496	2265	11·028504	39
22	970947	2238	998092	20	972855	2257	027145	38
23	972289	2231	998080	20	974209	2251	025791	37
24	973628	2224	998068	20	975560	2244	024440	36
25	974962	2217	998056	20	976906	2237	023004	35
26	976293	2210	998044	20	978248	2230	021752	34
27	977619	2203	998032	20	979586	2223	020414	33
28	978941	2197	998020	20	980021	2217	019070	32
29	980259	2190	998008	20	982251	2210	017749	31
30	981573	2183	997996	20	983577	2204	016423	30
31	8·982883	2177	9·997984	20	8·984899	2197	11·015101	29
32	984189	2170	997972	20	986217	2191	013783	28
33	985491	2163	997959	20	987532	2184	012468	27
34	986789	2157	997947	20	988842	2178	011158	26
35	988083	2150	997935	21	990149	2171	009851	25
36	989374	2144	997922	21	991451	2165	008549	24
37	990660	2138	997910	21	992750	2158	007250	23
38	991943	2131	997897	21	994045	2152	005955	22
39	993222	2125	997885	21	995337	2146	004663	21
40	994497	2119	997872	21	996624	2140	003376	20
41	8·995768	2112	9·997860	21	8·997908	2134	11·002092	19
42	997036	2106	997847	21	999188	2127	000812	18
43	998299	2100	997835	21	000465	2121	10·999535	17
44	999560	2094	997822	21	001738	2115	998262	16
45	9·000816	2087	997809	21	003007	2109	996993	15
46	002069	2082	997797	21	004272	2103	995728	14
47	003318	2076	997784	21	005534	2097	994466	13
48	004503	2070	997771	21	006792	2091	993208	12
49	005805	2064	997758	21	008047	2085	991953	11
50	007044	2058	997745	21	009298	2080	990702	10
51	8·008278	2052	8·997732	21	8·010546	2074	10·989454	9
52	009510	2046	997719	21	011790	2068	989210	8
53	010737	2040	997706	21	013031	2062	986969	7
54	011962	2034	997693	22	014268	2056	985732	6
55	013182	2029	997680	22	015502	2051	984498	5
56	014400	2023	997667	22	016732	2045	983268	4
57	015613	2017	997654	22	017959	2040	982041	3
58	016824	2012	997641	22	019183	2033	980817	2
59	018031	2006	997628	22	020403	2028	979597	1
60	019235	2000	997614	22	021620	2023	978380	0
	Cosine		Sine		Cotang.		Tang.	M.

84 Degrees.

(6 Degrees.) A TABLE OF LOGARITHMIC

M.	Sine	D.	Cosine	D.	Tang.	D.	Cotang.	
0	9·019235	2000	0·997614	22	9·021620	2023	10·978380	60
1	020435	1995	997601	22	022834	2017	977166	59
2	021632	1989	997589	22	024044	2011	975956	58
3	022825	1984	997574	22	025251	2006	974749	57
4	024016	1978	997561	22	026455	2000	973545	56
5	025203	1973	997547	22	027655	1995	972345	55
6	026386	1967	997534	23	028852	1990	971148	54
7	027567	1962	997520	23	030046	1985	969954	53
8	028744	1957	997507	23	031237	1979	968763	52
9	029918	1951	997493	23	032425	1974	967575	51
10	031089	1947	997480	23	033009	1969	966391	50
11	9·032257	1941	9·997466	23	9·034791	1964	10·965209	49
12	033421	1936	997452	23	035969	1958	964031	48
13	034582	1930	997439	23	037144	1953	962856	47
14	035741	1925	997425	23	038316	1948	961684	46
15	036896	1920	997411	23	039485	1943	960515	45
16	038048	1915	997397	23	040651	1938	959349	44
17	039197	1910	997383	23	041813	1933	958187	43
18	040342	1905	997369	23	042973	1928	957027	42
19	041485	1899	997355	23	044130	1923	955870	41
20	042625	1894	997341	23	045284	1918	954716	40
21	9·043760	1889	9·997327	24	9·046434	1913	10·953566	39
22	044895	1884	997313	24	047582	1908	952418	38
23	046026	1879	997299	24	048727	1903	951273	37
24	047154	1875	997285	24	049869	1898	950131	36
25	048279	1870	997271	24	051008	1893	948992	35
26	049400	1865	997257	24	052144	1889	947856	34
27	050519	1860	997242	24	053277	1884	946723	33
28	051635	1855	997228	24	054407	1879	945593	32
29	052749	1850	997214	24	055535	1874	944465	31
30	053859	1845	997199	24	056659	1870	943341	30
31	054966	1841	9·997185	24	9·057781	1865	10·942219	29
32	056071	1836	997170	24	058900	1860	941100	28
33	057172	1831	997156	24	060016	1855	939984	27
34	058271	1827	997141	24	061130	1851	938870	26
35	059367	1822	997127	24	062240	1846	937760	25
36	060460	1817	997112	24	063348	1842	936652	24
37	061551	1813	997098	24	064453	1837	935547	23
38	062639	1808	997083	25	065556	1833	934444	22
39	063724	1804	997068	25	066655	1828	933345	21
40	064806	1799	997053	25	067752	1824	932248	20
41	9·065885	1794	9·997039	25	9·068846	1819	10·931154	19
42	066962	1790	997024	25	069938	1815	930062	18
43	068036	1786	997009	25	071027	1810	928973	17
44	069107	1781	996994	25	072113	1806	927887	16
45	070176	1777	996979	25	073197	1802	926803	15
46	071242	1772	996964	25	074278	1797	925722	14
47	072306	1768	996949	25	075356	1793	924644	13
48	073366	1763	996934	25	076432	1789	923568	12
49	074424	1759	996919	25	077505	1784	922495	11
50	075480	1755	996904	25	078576	1780	921424	10
51	9·076533	1750	9·996889	25	9·079644	1776	10·920356	9
52	077583	1746	996874	25	080710	1772	919290	8
53	078631	1742	996858	25	081773	1767	918227	7
54	079676	1738	996843	25	082833	1763	917167	6
55	080719	1733	996828	25	083891	1759	916109	5
56	081759	1729	996812	26	084947	1755	915053	4
57	082797	1725	996797	26	086000	1751	914000	3
58	083832	1721	996782	26	087050	1747	912950	2
59	084864	1717	996766	26	088098	1743	911902	1
60	085894	1713	996751	26	089144	1738	910856	0
	Cosine		Sine		Cotang.		Tang.	M.

83 Degrees.

M	Sine	D.	Cosine	D.	Tang.	D.	Cotang.	
0	9·085894	1713	9·996751	26	9·089144	1738	10·910856	60
1	086922	1709	996735	26	090187	1734	909813	59
2	087947	1704	996720	26	091229	1730	908772	58
3	088970	1700	996704	26	092266	1727	907734	57
4	089990	1696	996688	26	093302	1723	906698	56
5	091008	1692	996673	26	094336	1719	905664	55
6	092024	1689	996657	26	095367	1715	904633	54
7	093037	1684	996641	26	096395	1711	903605	53
8	094047	1680	996625	26	097422	1707	902578	52
9	095056	1676	996610	26	098446	1703	901554	51
10	096062	1673	996504	26	099468	1699	900532	50
11	9·097065	1668	9·996578	27	9·100487	1695	10·899513	49
12	098066	1665	996562	27	101504	1691	898496	48
13	099065	1661	996546	27	102519	1687	897481	47
14	100062	1657	996530	27	103532	1684	896468	46
15	101056	1653	996514	27	104542	1680	895458	45
16	102048	1649	996498	27	105550	1676	894450	44
17	103037	1645	996482	27	106556	1672	893444	43
18	104025	1641	996465	27	107559	1669	892441	42
19	105010	1638	996449	27	108560	1665	891440	41
20	105992	1634	996433	27	109559	1661	890441	40
21	9·106973	1630	9·996417	27	9·110556	1658	10·889444	39
22	107951	1627	996400	27	111551	1654	888449	38
23	108927	1623	996384	27	112543	1650	887457	37
24	109901	1619	996368	27	113533	1646	886467	36
25	110873	1616	996351	27	114521	1643	885479	35
26	111842	1612	996335	27	115507	1639	884493	34
27	112809	1608	996318	27	116491	1636	883509	33
28	113774	1605	996302	28	117472	1632	882528	32
29	114737	1601	996285	28	118452	1629	881548	31
30	115698	1597	996269	28	119429	1625	880571	30
31	9·116656	1594	9·996252	28	9·120404	1622	10·879596	29
32	117613	1590	996235	28	121377	1618	878623	28
33	118567	1587	996219	28	122348	1615	877652	27
34	119519	1583	996202	28	123317	1611	876683	26
35	120460	1580	996185	28	124284	1607	875716	25
36	121417	1576	996168	28	125249	1604	874751	24
37	122362	1573	996151	28	126211	1601	873789	23
38	123306	1569	996134	28	127172	1597	872828	22
39	124248	1566	996117	28	128130	1594	871870	21
40	125187	1562	996100	28	129087	1591	870913	20
41	9·126125	1559	9·996083	29	9·130041	1587	10·869959	19
42	127060	1556	996066	29	130994	1584	869006	18
43	127993	1552	996049	29	131944	1581	868056	17
44	128925	1549	996032	29	132893	1577	867107	16
45	129854	1545	996015	29	133839	1574	866161	15
46	130781	1542	995998	29	134784	1571	865216	14
47	131706	1539	995980	29	135726	1567	864274	13
48	132630	1535	995963	29	136667	1564	863333	12
49	133551	1532	995946	29	137605	1561	862395	11
50	134470	1529	995928	29	138542	1558	861458	10
51	9·135387	1525	9·995911	29	9·139476	1555	10·860524	9
52	136303	1522	995894	29	140409	1551	859591	8
53	137216	1519	995876	29	141340	1548	858660	7
54	138128	1516	995859	29	142269	1545	857731	6
55	139037	1512	995841	29	143196	1542	856804	5
56	139944	1509	995823	29	144121	1539	855879	4
57	140850	1506	995806	29	145044	1535	854956	3
58	141754	1503	995788	29	145966	1532	854034	2
59	142655	1500	995771	29	146885	1529	853115	1
60	143555	1496	995753	29	147803	1526	852197	0
	Cosine		Sine		Cotang.		Tang.	M.

82 Degrees.

M.	Sine	D.	Cosine	D.	Tang.	D.	Cotang.	
0	9·143555	1496	9·995753	30	9·147803	1526	10·852197	60
1	144453	1493	995735	30	148718	1523	851282	59
2	145349	1490	995717	30	149632	1520	850368	58
3	146243	1487	995699	30	150544	1517	849456	57
4	147136	1484	995681	30	151454	1514	848546	56
5	148026	1481	995664	30	152363	1511	847637	55
6	148915	1478	995646	30	153269	1508	846731	54
7	149802	1475	995628	30	154174	1505	845826	53
8	150686	1472	995610	30	155077	1502	844923	52
9	151569	1469	995591	30	155978	1499	844022	51
10	152451	1466	995573	30	156877	1496	843123	50
11	9·153330	1463	0·995555	30	9·157775	1493	10·842225	49
12	154208	1460	995537	30	158671	1490	841329	48
13	155083	1457	995519	30	159565	1487	840435	47
14	155957	1454	995501	31	160457	1484	839543	46
15	156830	1451	995483	31	161347	1481	838653	45
16	157700	1448	995464	31	162236	1479	837764	44
17	158569	1445	995446	31	163123	1476	836877	43
18	159435	1442	995427	31	164008	1473	835992	42
19	160301	1439	995409	31	164892	1470	835108	41
20	161164	1436	995390	31	165774	1467	834226	40
21	9·162025	1433	9·995372	31	9·166654	1464	10·833346	39
22	162885	1430	995353	31	167539	1461	832468	38
23	163743	1427	995334	31	168409	1458	831591	37
24	164600	1424	995316	31	169284	1455	830716	36
25	165454	1422	995297	31	170157	1453	829843	35
26	166307	1419	995278	31	171029	1450	828971	34
27	167159	1416	995260	31	171899	1447	828101	33
28	168008	1413	995241	32	172767	1444	827233	32
29	168856	1410	995222	32	173634	1442	826366	31
30	169702	1407	995203	32	174499	1439	825501	30
31	9·170547	1405	9·995184	32	9·175362	1436	10·824638	29
32	171389	1402	995165	32	176224	1433	823776	28
33	172230	1399	995146	32	177084	1431	822916	27
34	173070	1396	995127	32	177942	1428	822058	26
35	173908	1394	995108	32	178799	1425	821201	25
36	174744	1391	995089	32	179655	1423	820345	24
37	175579	1388	995070	32	180508	1420	819492	23
38	176411	1386	995051	32	181360	1417	818640	22
39	177242	1383	995032	32	182211	1415	817789	21
40	178072	1380	995013	32	183059	1412	816941	20
41	9·178900	1377	9·994993	32	9·183907	1409	10·816093	19
42	179726	1374	994974	32	184752	1407	815248	18
43	180551	1372	994955	32	185597	1404	814403	17
44	181374	1369	994935	32	186439	1402	813561	16
45	182196	1366	994916	33	187280	1399	812720	15
46	183016	1364	994896	33	188120	1396	811880	14
47	183834	1361	994877	33	188958	1393	811042	13
48	184651	1359	994857	33	189794	1391	810206	12
49	185466	1356	994838	33	190629	1389	809371	11
50	186280	1353	994818	33	191462	1386	808538	10
51	9·187092	1351	9·994798	33	9·192294	1384	10·807706	9
52	187903	1348	994779	33	193124	1381	806876	8
53	188712	1346	994759	33	193953	1379	806047	7
54	189519	1343	994739	33	194780	1376	805220	6
55	190325	1341	994719	33	195606	1374	804394	5
56	191130	1338	994700	33	196430	1371	803570	4
57	191933	1336	994680	33	197253	1369	802747	3
58	192734	1333	994660	33	198074	1366	801926	2
59	193534	1330	994640	33	198894	1364	801106	1
60	194332	1328	994620	33	199713	1361	800287	0
	Cosine		Sine		Cotang.		Tang.	M.

81 Degrees.

M.	Sine	D.	Cosine	D.	Tang.	D.	Cotang.	
0	9·194332	1328	9·994680	33	9·199713	1361	10·800287	60
1	195129	1326	994600	33	200589	1359	799471	59
2	1C5925	1323	994580	33	201345	1356	798655	58
3	196719	1321	994560	34	202159	1354	797841	57
4	197511	1318	994540	34	202971	1352	797029	56
5	198302	1316	994519	34	203782	1349	796218	55
6	199091	1313	994499	34	204592	1347	795408	54
7	199879	1311	994479	34	205400	1345	794600	53
8	200666	1308	994459	34	206207	1342	793793	52
9	201451	1306	994438	34	207013	1340	792987	51
10	202234	1304	994418	34	207817	1338	792183	50
11	9·203017	1301	9·994397	34	9·208619	1335	10·791381	49
12	203797	1299	994377	34	209420	1333	790580	48
13	204577	1296	994357	34	210220	1331	789780	47
14	205354	1294	994336	34	211018	1328	788982	46
15	206131	1292	994316	34	211815	1326	788185	45
16	206906	1289	994295	34	212611	1324	787389	44
17	207679	1287	994274	35	213405	1321	786595	43
18	208452	1285	994254	35	214198	1319	785802	42
19	209222	1282	994233	35	214989	1317	785011	41
20	209992	1280	994212	35	215780	1315	784220	40
21	9·210760	1278	9·994191	35	9·216568	1312	10·783432	39
22	211526	1275	994171	35	217356	1310	782644	38
23	212291	1273	994150	35	218142	1308	781858	37
24	213055	1271	994129	35	218926	1305	781074	36
25	213818	1268	994108	35	219710	1303	780290	35
26	214579	1266	994087	35	220492	1301	779508	34
27	215338	1264	994066	35	221272	1299	778728	33
28	216097	1261	994045	35	222052	1297	777948	32
29	216854	1259	994024	35	222830	1294	777170	31
30	217609	1257	994003	35	223606	1292	776394	30
31	9·218363	1255	9·993981	35	9·224382	1290	10·775618	29
32	219116	1253	993960	35	225150	1288	774844	28
33	219868	1250	993939	35	225929	1286	774071	27
34	220618	1248	993918	35	226700	1284	773300	26
35	221367	1246	993896	36	227471	1281	772529	25
36	222115	1244	993875	36	228239	1279	771761	24
37	222861	1242	993854	36	229007	1277	770993	23
38	223606	1239	993832	36	229773	1275	770227	22
39	224349	1237	993811	36	230539	1273	769461	21
40	225092	1235	993789	36	231302	1271	768698	20
41	9·225833	1233	9·993768	36	9·232065	1269	10·767935	19
42	226573	1231	993746	36	232826	1267	767174	18
43	227311	1228	993725	36	233586	1265	766414	17
44	228048	1226	993703	36	234345	1263	765655	16
45	228784	1224	993681	36	235103	1260	764897	15
46	229518	1222	993660	36	235859	1258	764141	14
47	230252	1220	993638	36	236614	1256	763386	13
48	230984	1218	993616	36	237368	1254	762632	12
49	231714	1216	993594	37	238120	1252	761880	11
50	232444	1214	993572	37	238872	1250	761128	10
51	9·233172	1212	9·993550	37	9·239622	1248	10·760378	9
52	233899	1209	993528	37	240373	1246	759629	8
53	234625	1207	993506	37	241118	1244	758882	7
54	235349	1205	993484	37	241865	1242	758135	6
55	236073	1203	993462	37	242610	1240	757390	5
56	236795	1201	993440	37	243354	1238	756646	4
57	237515	1199	993418	37	244097	1236	755903	3
58	238235	1197	993396	37	244839	1234	755161	2
59	238953	1195	993374	37	245579	1232	754421	1
60	239670	1193	993351	37	246319	1230	753681	0
	Cosine		Sine		Cotang.		Tang.	M.

80 Degrees.

M.	Sine	D.	Cosine	D.	Tang.	D.	Cotang.	
0	9·239670	1193	9·993351	37	9·246319	1230	10·753681	60
1	240863	1191	993329	37	247057	1228	752943	59
2	241101	1189	993307	37	247794	1226	752206	58
3	241814	1187	993285	37	248530	1224	751470	57
4	242526	1185	993262	37	249264	1222	750736	56
5	243237	1183	993240	37	249998	1220	750002	55
6	243947	1181	993217	38	250730	1218	749270	54
7	244656	1179	993195	38	251461	1217	748539	53
8	245363	1177	993172	38	252191	1215	747809	52
9	246069	1175	993149	38	252920	1213	747080	51
10	246775	1173	993127	38	253648	1211	746352	50
11	9·247478	1171	9·993104	38	9·254374	1209	10·745626	49
12	248181	1169	993081	38	255100	1207	744900	48
13	248883	1167	993059	38	255824	1205	744176	47
14	249583	1165	993036	38	256547	1203	743453	46
15	250282	1163	993013	38	257269	1201	742731	45
16	250980	1161	992990	38	257990	1200	742010	44
17	251677	1159	992967	38	258710	1198	741290	43
18	252373	1158	992944	38	259429	1196	740571	42
19	253067	1156	992921	38	260146	1194	739854	41
20	253761	1154	992898	38	260863	1192	739137	40
21	9·254453	1152	9·992875	38	9·261578	1190	10·738422	39
22	255144	1150	992852	38	262292	1189	737708	38
23	255834	1148	992829	39	263005	1187	736995	37
24	256523	1146	992806	39	263717	1185	736283	36
25	257211	1144	992783	39	264428	1183	735572	35
26	257898	1142	992759	39	265138	1181	734862	34
27	258583	1141	992736	39	265847	1179	734153	33
28	259268	1139	992713	39	266555	1178	733445	32
29	259951	1137	992690	39	267261	1176	732739	31
30	260633	1135	992666	39	267967	1174	732033	30
31	9·261314	1133	9·992643	39	9·268671	1172	10·731329	29
32	261994	1131	992619	39	269375	1170	730625	28
33	262673	1130	992596	39	270077	1169	729923	27
34	263351	1128	992572	39	270779	1167	729221	26
35	264027	1126	992549	39	271479	1165	728521	25
36	264703	1124	992525	39	272178	1164	727822	24
37	265377	1122	992501	39	272876	1162	727124	23
38	266051	1120	992478	40	273573	1160	726427	22
39	266723	1119	992454	40	274269	1158	725731	21
40	267395	1117	992430	40	274964	1157	725036	20
41	9·268065	1115	9·992406	40	9·275658	1155	10·724342	19
42	268734	1113	992382	40	276351	1153	723649	18
43	269402	1111	992359	40	277043	1151	722957	17
44	270069	1110	992335	40	277734	1150	722266	16
45	270735	1108	992311	40	278424	1148	721576	15
46	271400	1106	992287	40	279113	1147	720887	14
47	272064	1105	992263	40	279801	1145	720199	13
48	272726	1103	992239	40	280488	1143	719512	12
49	273388	1101	992214	40	281174	1141	718826	11
50	274049	1099	992190	40	281858	1140	718142	10
51	9·274708	1098	9·992166	40	9·282542	1138	10·717458	9
52	275367	1096	992142	40	283225	1136	716775	8
53	276024	1094	992117	41	283907	1135	716093	7
54	276681	1092	992093	41	284588	1133	715412	6
55	277337	1091	992069	41	285268	1131	714732	5
56	277991	1089	992044	41	285947	1130	714053	4
57	278644	1087	992020	41	286624	1128	713376	3
58	279297	1086	991996	41	287301	1126	712699	2
59	279948	1084	991971	41	287977	1125	712023	1
60	280599	1082	991947	41	288652	1123	711348	0
	Cosine		Sine		Cotang.		Tang.	M.

70 Degrees.

M.	Sine	D.	Cosine	D.	Tang.	D.	Cotang.	
0	9·283050	1082	9·991947	41	9·288652	1123	10·711348	60
1	281248	1081	991922	41	289326	1122	710674	59
2	281897	1079	991897	41	289999	1120	710001	58
3	282544	1077	991873	41	290671	1118	709329	57
4	283190	1076	991848	41	291342	1117	708658	56
5	283836	1074	991823	41	292013	1115	707987	55
6	284480	1072	991799	41	292682	1114	707318	54
7	285124	1071	991774	42	293350	1112	706650	53
8	285766	1069	991749	42	294017	1111	705983	52
9	286408	1067	991724	42	294684	1109	705316	51
10	287048	1066	991699	42	295349	1107	704651	50
11	9·287687	1064	9·991674	42	9·296013	1106	10·703987	49
12	288326	1063	991649	42	296677	1104	703323	48
13	288964	1061	991624	42	297339	1103	702661	47
14	289600	1059	991599	42	298001	1101	701999	46
15	290236	1058	991574	42	298662	1100	701338	45
16	290870	1056	991549	42	299322	1098	700678	44
17	291504	1054	991524	42	299980	1096	700020	43
18	292137	1053	991498	42	300638	1095	699362	42
19	292768	1051	991473	42	301295	1093	698705	41
20	293399	1050	991448	42	301951	1092	698049	40
21	9·294029	1048	9·991422	42	9·302607	1090	10·697393	39
22	294658	1046	991397	42	303261	1089	696739	38
23	295286	1045	991372	43	303914	1087	696086	37
24	295913	1043	991346	43	304567	1086	695433	36
25	296539	1042	991321	43	305218	1084	694782	35
26	297164	1040	991295	43	305869	1083	694131	34
27	297788	1039	991270	43	306519	1081	693481	33
28	298412	1037	991244	43	307168	1080	692832	32
29	299034	1036	991218	43	307815	1078	692185	31
30	299655	1034	991103	43	308463	1077	691537	30
31	9·300276	1032	9·991167	43	9·309109	1075	10·690891	29
32	300895	1031	991141	43	309754	1074	690246	28
33	301514	1029	991115	43	310398	1073	689602	27
34	302132	1028	991090	43	311042	1071	688958	26
35	302748	1026	991064	43	311685	1070	688315	25
36	303364	1025	991038	43	312327	1068	687673	24
37	303979	1023	991012	43	312967	1067	687033	23
38	304593	1022	990986	43	313608	1065	686392	22
39	305207	1020	990960	43	314247	1064	685753	21
40	305819	1019	990934	44	314885	1062	685115	20
41	9·306430	1017	9·990908	44	9·315523	1061	10·684477	19
42	307041	1016	990882	44	316159	1060	683841	18
43	307650	1014	990855	44	316795	1058	683205	17
44	308259	1013	990829	44	317430	1057	682570	16
45	308867	1011	990803	44	318064	1055	681936	15
46	309474	1010	990777	44	318697	1054	681303	14
47	310080	1008	990750	44	319329	1053	680671	13
48	310685	1007	990724	44	319961	1051	680039	12
49	311289	1005	990697	44	320592	1050	679408	11
50	311893	1004	990671	44	321222	1048	678778	10
51	9·312495	1003	9·990644	44	9·321851	1047	10·678149	9
52	313097	1001	990618	44	322479	1045	677521	8
53	313698	1000	990591	44	323106	1044	676894	7
54	314297	998	990565	44	323733	1043	676267	6
55	314897	997	990538	44	324358	1041	675642	5
56	315495	996	990511	45	324983	1040	675017	4
57	316092	994	990485	45	325607	1039	674393	3
58	316689	993	990458	45	326231	1037	673769	2
59	317284	991	990431	45	326853	1036	673147	1
60	317879	990	990404	45	327475	1035	672525	0
	Cosine		Sine		Cotang.		Tang.	M.

78 Degrees.

M.	Sine	D.	Cosine	D.	Tang.	D.	Cotang.	
0	9·317879	990	9·990404	45	9·327474	1035	10·672526	60
1	318473	988	990378	45	328095	1033	671905	59
2	319066	987	990351	45	328715	1032	671285	58
3	319658	986	990324	45	329334	1030	670666	57
4	320249	984	990297	45	329953	1029	670047	56
5	320840	983	990270	45	330570	1028	669430	55
6	321430	982	990243	45	331187	1026	668813	54
7	322019	980	990215	45	331803	1025	668197	53
8	322607	979	990188	45	332418	1024	667582	52
9	323194	977	990161	45	333033	1023	666967	51
10	323780	976	990134	45	333646	1021	666354	50
11	9·324366	975	9·990107	46	9·334259	1020	10·665741	49
12	324950	973	990079	46	334871	1019	665129	48
13	325534	972	990052	46	335482	1017	664518	47
14	326117	970	990025	46	336093	1016	663907	46
15	326700	969	989997	46	336702	1015	663298	45
16	327281	968	989970	46	337311	1013	662689	44
17	327862	966	989942	46	337919	1012	662081	43
18	328442	965	989915	46	338527	1011	661473	42
19	329021	964	989887	46	339133	1010	660867	41
20	329599	962	989860	46	339739	1008	660261	40
21	9·330176	961	9·989832	46	9·340344	1007	10·659656	39
22	330753	960	989804	46	340948	1006	659052	38
23	331329	958	989777	46	341552	1004	658448	37
24	331903	957	989749	47	342155	1003	657845	36
25	332478	956	989721	47	342757	1002	657243	35
26	333051	954	989693	47	343358	1000	656642	34
27	333624	953	989665	47	343958	999	656042	33
28	334105	952	989637	47	344558	998	655442	32
29	334766	950	989609	47	345157	997	654843	31
30	335337	949	989582	47	345755	996	654245	30
31	9·335906	948	9·989553	47	9·346353	994	10·653647	29
32	336475	946	989525	47	346949	993	653051	28
33	337043	945	989497	47	347545	992	652455	27
34	337610	944	989469	47	348141	991	651859	26
35	338176	943	989441	47	348735	990	651265	25
36	338742	941	989413	47	349329	988	650671	24
37	339306	940	989384	47	349922	987	650078	23
38	339871	939	989356	47	350514	986	649486	22
39	340434	937	989328	47	351106	985	648894	21
40	340996	936	989300	47	351697	983	648303	20
41	9·341558	935	9·989271	47	9·352287	982	10·647713	19
42	342119	934	989243	47	352876	981	647124	18
43	342679	932	989214	47	353465	980	646535	17
44	343239	931	989186	47	354053	979	645947	16
45	343797	930	989157	47	354640	977	645360	15
46	344355	929	989128	48	355227	976	644773	14
47	344912	927	989100	48	355813	975	644187	13
48	345469	926	989071	48	356398	974	643602	12
49	346024	925	989042	48	356982	973	643018	11
50	346579	924	989014	48	357566	971	642434	10
51	9·347134	922	9·988985	48	9·358140	970	10·641851	9
52	347687	921	988956	48	358731	969	641269	8
53	348240	920	988927	48	359313	968	640687	7
54	348792	919	988898	48	359893	967	640107	6
55	349343	917	988869	48	360474	966	639526	5
56	349893	916	988840	48	361053	965	638947	4
57	350443	915	988811	49	361632	963	638368	3
58	350992	914	988782	49	362210	962	637790	2
59	351540	913	988753	49	362787	961	637213	1
60	352088	911	988724	49	363364	960	636636	0
	Cosine		Sine		Cotang.		Tang.	M.

M.	Sine	D.	Cosine	D.	Tang.	D.	Cotang.	
0	9·352088	911	9·988724	49	9·363364	960	10·636636	60
1	352635	910	988695	49	363940	959	636060	59
2	353181	909	988666	49	364515	958	635485	58
3	353726	908	988636	49	365090	957	634910	57
4	354271	907	988607	49	365664	955	634336	56
5	354815	905	988578	49	366237	954	633763	55
6	355358	904	988548	49	366810	953	633190	54
7	355901	903	988519	49	367382	952	632618	53
8	356443	902	988489	49	367953	951	632047	52
9	356984	901	988460	49	368524	950	631476	51
10	357524	899	988430	49	369094	949	630906	50
11	9·358064	898	9·988401	49	9·369663	948	10·630337	49
12	358603	897	988371	49	370232	946	629768	48
13	359141	896	988342	49	370799	945	629201	47
14	359678	895	988312	50	371307	944	628633	46
15	360215	893	988282	50	371933	943	628067	45
16	360752	892	988252	50	372499	942	627501	44
17	361287	891	988223	50	373064	941	626936	43
18	361822	890	988193	50	373629	940	626371	42
19	362356	889	988163	50	374193	939	625807	41
20	362889	888	988133	50	374756	938	625244	40
21	9·363422	887	9·988103	50	9·375319	937	10·624681	39
22	363954	885	988073	50	375881	935	624119	38
23	364485	884	988043	50	376442	934	623558	37
24	365016	883	988013	50	377003	933	622997	36
25	365546	882	987983	50	377563	939	622437	35
26	366075	881	987953	50	378122	931	621878	34
27	366604	880	987922	50	378681	930	621319	33
28	367131	879	987892	50	379239	929	620761	32
29	367659	877	987862	50	379797	928	620203	31
30	368185	876	987832	51	380354	927	619646	30
31	9·368711	875	9·987801	51	9·380910	926	10·619090	29
32	369236	874	987771	51	381466	925	618534	28
33	369761	873	987740	51	382020	924	617980	27
34	370285	872	987710	51	382575	923	617425	26
35	370808	871	987679	51	383129	922	616871	25
36	371330	870	987649	51	383682	921	616318	24
37	371852	869	987618	51	384234	920	615766	23
38	372373	867	987588	51	384786	919	615214	22
39	372894	866	987557	51	385337	918	614663	21
40	373414	865	987526	51	385888	917	614112	20
41	9·373933	864	9·987496	51	9·386438	915	10·613562	19
42	374452	863	987465	51	386987	914	613013	18
43	374970	862	987434	51	387536	913	612464	17
44	375487	861	987403	52	388084	912	611916	16
45	376003	860	987372	52	388631	911	611369	15
46	376519	859	987341	52	389178	910	610822	14
47	377035	858	987310	52	389724	909	610276	13
48	377549	857	987279	52	390270	908	609730	12
49	378063	856	987248	52	390815	907	609185	11
50	378577	854	987217	52	391360	906	608640	10
51	9·379089	853	9·987186	52	9·391903	905	10·608097	9
52	379601	852	987155	52	392447	904	607553	8
53	380113	851	987124	52	392989	903	607011	7
54	380624	850	987092	52	393531	902	606469	6
55	381134	849	987061	52	394073	901	605927	5
56	381643	848	987030	52	394614	900	605386	4
57	382152	847	986998	52	395154	899	604846	3
58	382661	846	986967	52	395694	898	604306	2
59	383168	845	986936	52	396233	897	603767	1
60	383675	844	986904	52	396771	896	603229	0
	Cosine		Sine		Cotang.		Tang.	M.

76 Degrees.

M.	Sine	D.	Cosine	D.	Tang.	D.	Cotang.	
0	9·383675	844	9·986904	52	9·396771	896	10·603229	60
1	384162	843	986873	53	397309	896	602691	59
2	384687	842	986841	53	397846	895	602154	58
3	385192	841	986809	53	398383	894	601617	57
4	385697	840	986778	53	398919	893	601081	56
5	386201	839	986746	53	399455	892	600545	55
6	386704	838	986714	53	399990	891	600010	54
7	387207	837	986683	53	400524	890	599476	53
8	387709	836	986651	53	401058	889	598942	52
9	388210	835	986619	53	401591	888	598409	51
10	388711	834	986587	53	402124	887	597876	50
11	9·389211	833	9·986555	53	9·402656	886	10·597344	49
12	389711	832	986523	53	403187	885	596813	48
13	390210	831	986491	53	403718	884	596282	47
14	390708	830	986459	53	404249	883	505751	46
15	391206	829	986427	53	404778	882	505222	45
16	391703	827	986395	53	405308	881	504692	44
17	392199	826	986363	54	405836	880	594164	43
18	392695	825	986331	54	406364	879	593636	42
19	393191	824	986299	54	406892	878	593108	41
20	393685	823	986266	54	407419	877	592581	40
21	9·394179	822	9·986234	54	9·407945	876	10·592055	39
22	394673	821	986202	54	408471	875	591529	38
23	395166	820	986169	54	408997	874	591003	37
24	395658	819	986137	54	409521	874	590479	36
25	396150	818	986104	54	410045	873	589955	35
26	396641	817	986072	54	410569	872	589431	34
27	397132	817	986039	54	411092	871	588908	33
28	397621	816	986007	54	411615	870	588385	32
29	398111	815	985974	54	412137	869	587863	31
30	398600	814	985942	54	412658	868	587342	30
31	9·399088	813	9·985909	55	9·413179	867	10·586821	29
32	399575	812	985876	55	413699	866	586301	28
33	400062	811	985843	55	414219	865	585781	27
34	400549	810	985811	55	414738	864	585262	26
35	401035	809	985778	55	415257	864	584743	25
36	401520	808	985745	55	415775	863	584225	24
37	402005	807	985712	55	416293	862	583707	23
38	402489	806	985679	55	416810	861	583190	22
39	402972	805	985646	55	417326	860	582674	21
40	403455	804	985613	55	417842	859	582158	20
41	9·403938	803	9·985580	55	9·418358	858	10·581642	19
42	404420	802	985547	55	418873	857	581127	18
43	404901	801	985514	55	419387	856	580613	17
44	405382	800	985480	55	419901	855	580099	16
45	405862	799	985447	55	420415	855	579585	15
46	406341	798	985414	56	420927	854	579073	14
47	406820	797	985380	56	421440	853	578560	13
48	407299	796	985347	56	421952	852	578048	12
49	407777	795	985314	56	422463	851	577537	11
50	408254	794	985280	56	422974	850	577026	10
51	9·408731	794	9·985247	56	9·423484	849	10·576516	9
52	409207	793	985213	56	423993	848	576007	8
53	409682	792	985180	56	424503	848	575497	7
54	410157	791	985146	56	425011	847	574989	6
55	410632	790	985113	56	425519	846	574481	5
56	411106	789	985079	56	426027	845	573973	4
57	411579	788	985045	56	426534	844	573466	3
58	412052	787	985011	56	427041	843	572959	2
59	412524	786	984978	56	427547	843	572453	1
60	412996	785	984944	56	428052	842	571948	0
	Cosine		Sine		Cotang.		Tang.	M.

75 Degrees.

M.	Sine	D.	Cosine	D.	Tang.	D.	Cotang.	
0	0·412996	785	9·984944	57	9·428052	842	10·571948	60
1	413467	784	984910	57	428557	841	571443	59
2	413938	783	984876	57	429062	840	570938	58
3	414408	783	984842	57	429566	839	570434	57
4	414878	782	984808	57	430070	838	569930	56
5	415347	781	984774	57	430573	838	569427	55
6	415815	780	984740	57	431075	837	568925	54
7	416283	779	984706	57	431577	836	568423	53
8	416751	778	984672	57	432079	835	567921	52
9	417217	777	984637	57	432580	834	567420	51
10	417684	776	984603	57	433080	833	566920	50
11	0·418150	775	0·984569	57	9·433580	832	10·566420	49
12	418615	774	984535	57	434080	832	565920	48
13	419079	773	984500	57	434579	831	565421	47
14	419544	773	984466	57	435078	830	564922	46
15	420007	772	984432	58	435576	829	564424	45
16	420470	771	984397	58	436073	828	563927	44
17	420933	770	984363	58	436570	828	563430	43
18	421395	769	984328	58	437067	827	562933	42
19	421857	768	984294	58	437563	826	562437	41
20	422318	767	984259	58	438059	825	561941	40
21	9·422778	767	9·984224	58	9·438554	824	10·561446	39
22	423238	766	984190	58	439048	823	560952	38
23	423697	765	984155	58	439543	823	560457	37
24	424156	764	984120	58	440036	822	559964	36
25	424615	763	984085	58	440529	821	559471	35
26	425073	762	984050	58	441022	820	558978	34
27	425530	761	984015	58	441514	819	558486	33
28	425987	760	983981	58	442006	819	557994	32
29	426443	760	983946	58	442497	818	557503	31
30	426899	759	983911	58	442988	817	557012	30
31	9·427354	758	9·983875	58	9·443479	816	10·556521	29
32	427809	757	983840	59	443968	816	556032	28
33	428263	756	983805	59	444458	815	555542	27
34	428717	755	983770	59	444947	814	555053	26
35	429170	754	983735	59	445435	813	554565	25
36	429623	753	983700	59	445923	812	554077	24
37	430075	752	983664	59	446411	812	553589	23
38	430527	752	983629	59	446898	811	553102	22
39	430978	751	983594	59	447384	810	552616	21
40	431429	750	983558	59	447870	809	552130	20
41	9·431879	749	9·983523	59	9·448356	809	10·551644	19
42	432329	749	983487	59	448841	808	551159	18
43	432778	748	983452	59	449326	807	550674	17
44	433226	747	983416	59	449810	806	550190	16
45	433675	746	983381	59	450294	806	549706	15
46	434122	745	983345	59	450777	805	549223	14
47	434569	744	983309	59	451260	804	548740	13
48	435016	744	983273	60	451743	803	548257	12
49	435462	743	983238	60	452225	802	547775	11
50	435908	742	983202	60	452706	802	547294	10
51	9·436353	741	9·983166	60	9·453187	801	10·546813	9
52	436798	740	983130	60	453668	800	546332	8
53	437242	740	983094	60	454148	799	545852	7
54	437686	739	983058	60	454628	799	545372	6
55	438129	738	983022	60	455107	798	544893	5
56	438572	737	982986	60	455586	797	544414	4
57	439014	736	982950	60	456064	796	543936	3
58	439456	736	982914	60	456542	796	543458	2
59	439897	735	982878	60	457019	795	542981	1
60	440338	734	982842	60	457496	794	542504	0
	Cosine		Sine		Cotang.		Tang.	M.

74 Degrees.

M.	Sine	D.	Cosne	D.	Tang.	D.	Cotang.	
0	9·440338	734	9·982842	60	9·457496	794	10·542504	60
1	440778	733	982805	60	457973	793	542027	59
2	441218	732	982769	61	458449	793	541551	58
3	441658	731	982733	61	458925	792	541075	57
4	442096	731	982696	61	459400	791	540600	56
5	442535	730	982660	61	459875	790	540125	55
6	442973	729	982624	61	460349	790	539651	54
7	443410	728	982587	61	460823	789	539177	53
8	443847	727	982551	61	461297	788	538703	52
9	444284	727	982514	61	461770	788	538230	51
10	444720	726	982477	61	462243	787	537758	50
11	9·445155	725	9·982441	61	9·462714	786	10·537286	49
12	445590	734	982404	61	463186	785	536814	48
13	446025	723	982367	61	463658	785	536342	47
14	446459	723	982331	61	464129	784	535871	46
15	446893	722	982294	61	464599	783	535401	45
16	447326	721	982257	61	465069	783	534931	44
17	447759	720	982220	62	465530	782	534461	43
18	448191	720	982183	62	466008	781	533992	42
19	448623	719	982146	62	466476	780	533524	41
20	449054	718	982109	62	466945	780	533055	40
21	9·449485	717	9·982072	62	9·467413	779	10·532587	39
22	449915	716	982035	62	467880	778	532120	38
23	450345	716	981998	62	468347	778	531653	37
24	450775	715	981961	62	468814	777	531186	36
25	451204	714	981924	62	469280	776	530720	35
26	451632	713	981886	62	469746	775	530254	34
27	452060	713	981849	62	470211	775	529789	33
28	452488	712	981812	62	470676	774	529324	32
29	452915	711	981774	62	471141	773	528859	31
30	453342	710	981737	62	471605	773	528395	30
31	9·453768	710	9·981699	63	9·472068	772	10·527932	29
32	454194	709	981662	63	472532	771	527468	28
33	454619	708	981625	63	472995	771	527005	27
34	455044	707	981587	63	473457	770	526543	26
35	455469	707	981549	63	473919	769	526081	25
36	455893	706	081512	63	474381	769	525619	24
37	456316	705	981474	63	474842	768	525158	23
38	456739	704	981436	63	475303	767	524697	22
39	457162	704	981399	63	475763	767	524237	21
40	457584	703	981361	63	476223	766	523777	20
41	9·458006	702	9·981323	63	9·476683	765	10·523317	19
42	458427	701	981285	63	477142	765	522858	18
43	458848	701	981247	63	477601	764	522399	17
44	459268	700	981209	63	478059	763	521941	16
45	459688	699	981171	63	478517	763	521483	15
46	460108	698	981133	64	478975	762	521025	14
47	460527	698	981095	64	479432	761	520568	13
48	460946	697	981057	64	479889	761	520111	12
49	461364	696	981019	64	480345	760	519655	11
50	461782	695	980981	64	480801	759	519199	10
51	9·462199	695	9·980942	64	9·481257	759	10·518743	9
52	462616	694	980904	64	481712	758	518288	8
53	463032	693	980866	64	482167	757	517833	7
54	463448	693	980827	64	482621	757	517379	6
55	463864	692	980789	64	483075	756	516925	5
56	464279	691	980750	64	483529	755	516471	4
57	464694	690	980712	64	483982	755	516018	3
58	465108	690	980673	64	484435	754	515565	2
59	465522	689	980635	64	484887	753	515113	1
60	465935	688	980596	64	485339	753	514661	0

| | Cosine | | Sine | | Cotang. | | Tang. | M. |

M.	Sine	D.	Cosine	D.	Tang.	D.	Cotang.	
0	9·465935	688	9·980506	64	9·485339	755	10·514661	60
1	466348	688	980558	64	485791	752	514209	59
2	466761	687	980519	65	486242	751	513758	58
3	467173	686	980480	65	486693	751	513307	57
4	467585	685	980442	65	487143	750	512857	56
5	467996	685	980403	65	487593	749	512407	55
6	468407	684	980364	65	488043	749	511957	54
7	468817	683	980325	65	488492	748	511508	53
8	469227	683	980296	65	488941	747	511059	52
9	469637	682	980247	65	489390	747	510610	51
10	470046	681	980208	65	489838	746	510162	50
11	9·470455	680	9·980169	65	9·490286	746	10·509714	49
12	470863	680	980130	65	490733	745	509267	48
13	471271	679	980091	65	491180	744	508820	47
14	471679	678	980052	65	491627	744	508373	46
15	472086	678	980012	65	492073	743	507927	45
16	472492	677	979973	65	492519	743	507481	44
17	472898	676	979934	66	492965	742	507035	43
18	473304	676	979895	66	493410	741	506590	42
19	473710	675	979855	66	493854	740	506146	41
20	474115	674	979816	66	494299	740	505701	40
21	9·474519	674	9·979776	66	9·494743	740	10·505257	39
22	474923	673	979737	66	495186	739	504814	38
23	475327	672	979697	66	495630	738	504370	37
24	475730	672	979658	66	496073	737	503927	36
25	476133	671	979618	66	496515	737	503485	35
26	476536	670	979579	66	496957	736	503043	34
27	476938	669	979539	66	497399	736	502601	33
28	477340	669	979499	66	497841	735	502159	32
29	477741	668	979459	66	498282	734	501718	31
30	478142	667	979420	66	498722	734	501278	30
31	9·478542	667	9·979380	66	9·499163	733	10·500837	29
32	478942	666	979340	66	499603	733	500397	28
33	479342	665	979300	67	500042	732	499958	27
34	479741	665	979260	67	500481	731	499519	26
35	480140	664	979220	67	500920	731	499080	25
36	480539	663	979180	67	501359	730	498641	24
37	480937	663	979140	67	501797	730	498203	23
38	481334	662	979100	67	502235	729	497765	22
39	481731	661	979059	67	502672	728	497328	21
40	482128	661	979019	67	503109	728	496891	20
41	9·482525	660	9·978979	67	9·503546	727	10·496454	19
42	482921	659	978939	67	503982	727	496018	18
43	483316	659	978898	67	504418	726	495582	17
44	483712	658	978858	67	504854	725	495146	16
45	484107	657	978817	67	505289	725	494711	15
46	484501	657	978777	67	505724	724	494276	14
47	484895	656	978736	67	506159	724	493841	13
48	485289	655	978696	68	506593	723	493407	12
49	485682	655	978655	68	507027	722	492973	11
50	486075	654	978615	68	507460	722	492540	10
51	9·486467	653	9·978574	68	9·507893	721	10·492107	9
52	486860	653	978533	68	508326	721	491674	8
53	487251	652	978493	68	508759	720	491241	7
54	487643	651	978452	68	509191	719	490809	6
55	488034	651	978411	68	509622	719	490378	5
56	488424	650	978370	68	510054	718	489946	4
57	488814	650	978329	68	510485	718	489515	3
58	489204	649	978288	68	510916	717	489084	2
59	489593	648	978247	68	511346	716	488654	1
60	489982	648	978206	68	511776	716	488224	0
	Cosine		Sine		Cotang.		Tang.	M.

72 Degrees

M.	Sine	D.	Cosine	D.	Tang.	D.	Cotang.	
0	9·489982	648	9·978206	68	9·511776	716	10·488224	60
1	490371	648	978165	68	512206	716	487794	59
2	490759	647	978124	68	512635	715	487365	58
3	491147	646	978083	69	513064	714	486936	57
4	491535	646	978042	69	513493	714	486507	56
5	491922	645	978001	69	513921	713	486079	55
6	492308	644	977959	69	514349	713	485651	54
7	492695	644	977918	69	514777	712	485223	53
8	493081	643	977877	69	515204	712	484796	52
9	493466	642	977835	69	515631	711	484369	51
10	493851	642	977794	69	516057	710	483943	50
11	9 494236	641	9·977752	69	9·516484	710	10·483516	49
12	494621	641	977711	69	516910	709	483090	48
13	495005	640	977669	69	517335	709	482665	47
14	495388	639	977628	69	517761	708	482239	46
15	495772	639	977586	69	518185	708	481815	45
16	496154	638	977544	70	518610	707	481390	44
17	496537	637	977503	70	519034	706	480966	43
18	496919	637	977461	70	519458	706	480542	42
19	497301	636	977419	70	519882	705	480118	41
20	497682	636	977377	70	520305	705	479695	40
21	9·498064	635	9·977335	70	9·520728	704	10·479272	39
22	498444	634	977293	70	521151	703	478849	38
23	498825	634	977251	70	521573	703	478427	37
24	499204	633	977209	70	521995	703	478005	36
25	499584	632	977167	70	522417	702	477583	35
26	499963	632	977125	70	522838	702	477162	34
27	500342	631	977083	70	523259	701	476741	33
28	500721	631	977041	70	523680	701	476320	32
29	501099	630	976999	70	524100	700	475900	31
30	501476	629	976957	70	524520	699	475480	30
31	9·501854	629	9·976914	70	9·524939	699	10·475061	29
32	502231	628	976872	71	525359	698	474641	28
33	502607	628	976830	71	525778	698	474222	27
34	502984	627	976787	71	526197	697	473803	26
35	503360	626	976745	71	526615	697	473385	25
36	503735	626	976702	71	527033	696	472967	24
37	504110	625	976660	71	527451	696	472549	23
38	504485	625	976617	71	527868	695	472132	22
39	504860	624	976574	71	528285	695	471715	21
40	505234	623	976532	71	528702	694	471298	20
41	9·505608	623	9·976489	71	9·529119	693	10·470881	19
42	505981	622	976446	71	529535	693	470465	18
43	506354	622	976404	71	529950	693	470050	17
44	506727	621	976361	71	530366	692	469634	16
45	507099	620	976318	71	530781	691	469219	15
46	507471	620	976275	71	531196	691	468804	14
47	507843	619	976232	72	531611	690	468389	13
48	508214	619	976189	72	532025	690	467975	12
49	508585	618	976146	72	532439	689	467561	11
50	508956	618	976103	72	532853	689	467147	10
51	9·509326	617	9·976060	72	9·533266	688	10·466734	9
52	509696	616	976017	72	533679	688	466321	8
53	510065	616	975974	72	534092	687	465908	7
54	510434	615	975930	72	534504	687	465496	6
55	510803	615	975887	72	534916	686	465084	5
56	511172	614	975844	72	535329	686	464672	4
57	511540	613	975800	72	535739	685	464261	3
58	511907	613	975757	72	536150	685	463850	2
59	512275	612	975714	72	536561	684	463439	1
60	512642	612	975670	72	536972	684	463028	0
	Cosine		Sine		Cotang.		Tang.	M.

71 Degrees.

M.	Sine	D.	Cosine	D.	Tang.	D.	Cotang.	
0	9·512642	612	9·975670	73	9·536972	684	10·463028	60
1	513009	611	975627	73	537382	683	462618	59
2	513375	611	975583	73	537792	683	462208	58
3	513741	610	975539	73	538202	682	461798	57
4	514107	609	975496	73	538611	682	461389	56
5	514472	609	975452	73	539020	681	460980	55
6	514837	608	975408	73	539429	681	460571	54
7	515202	608	975365	73	539837	680	460163	53
8	515566	607	975321	73	540245	680	459755	52
9	515930	607	975277	73	540653	679	459347	51
10	516294	606	975233	73	541061	679	458939	50
11	9·516657	605	9·975189	73	9·541468	678	10·458532	49
12	517020	605	975145	73	541875	678	458125	48
13	517382	604	975101	73	542281	677	457719	47
14	517745	604	975057	73	542688	677	457312	46
15	518107	603	975013	73	543094	676	456906	45
16	518468	603	974969	74	543499	676	456501	44
17	518829	602	974925	74	543905	675	456095	43
18	519190	601	974880	74	544310	675	455690	42
19	519551	601	974836	74	544715	674	455285	41
20	519911	600	974792	74	545119	674	454881	40
21	9·520271	600	9·974748	74	9·545524	673	10·454476	39
22	520631	599	974703	74	545928	673	454072	38
23	520990	599	974659	74	546331	672	453669	37
24	521349	598	974614	74	546735	672	453265	36
25	521707	598	974570	74	547138	671	452862	35
26	522066	597	974525	74	547540	671	452460	34
27	522424	596	974481	74	547943	670	· 452057	33
28	522781	596	974430	74	548345	670	451655	32
29	523138	595	974391	74	548747	669	451253	31
30	523495	595	974347	75	549149	669	450851	30
31	9·523852	594	9·974302	75	9·549550	668	10·450450	29
32	524208	594	974257	75	549951	668	450049	28
33	524564	593	974212	75	550352	667	449648	27
34	524920	593	974167	75	550752	667	449248	26
35	525275	592	974122	75	551152	666	448848	25
36	525630	591	974077	75	551552	666	448448	24
37	525984	591	974032	75	551952	665	448048	23
38	526339	590	973987	75	552351	665	447649	22
39	526693	590	973942	75	552750	665	447250	21
40	527046	589	973897	75	553149	664	446851	20
41	9·527400	589	9·973852	75	9·553548	664	10·446452	19
42	527753	588	973807	75	553946	663	446054	18
43	528105	588	973761	75	554344	663	445656	17
44	528458	587	973716	76	554741	662	445259	16
45	528810	587	973671	76	555139	662	444861	15
46	529161	586	973625	76	555536	661	444464	14
47	529513	586	973580	76	555933	661	444067	13
48	529864	585	973535	76	556329	660	443671	12
49	530215	585	973489	76	556725	660	443275	11
50	530565	584	973444	76	557121	659	442879	10
51	9·530915	584	9·973398	76	9·557517	659	10·442483	9
52	531265	583	973352	76	557913	659	442087	8
53	531614	582	973307	76	558308	658	441692	7
54	531963	582	973261	76	558702	658	441298	6
55	532312	581	973215	76	559097	657	440903	5
56	532661	581	973169	76	559491	657	440509	4
57	533009	580	973124	76	559885	656	440115	3
58	533357	580	973078	76	560279	656	439721	2
59	533704	579	973032	77	560673	655	439327	1
60	534052	578	972986	77	561066	655	438934	0
	Cosine		Sine		Cotang.		Tang.	M.

70 Degrees.

M.	Sine	D.	Cosine	D.	Tang.	D.	Cotang.	
0	9·534052	578	9·972986	77	9·561066	655	10·438934	60
1	534399	577	972940	77	561459	654	438541	59
2	534745	577	972894	77	561851	654	438149	58
3	535092	577	972848	77	562244	653	437756	57
4	535438	576	972802	77	562636	653	437364	56
5	535783	576	972755	77	563029	653	436971	55
6	536129	575	972709	77	563419	652	436581	54
7	536474	574	972663	77	563811	652	436189	53
8	536818	574	972617	77	564202	651	435798	52
9	537163	573	972570	77	564502	651	435408	51
10	537507	573	972524	77	564983	650	435017	50
11	9·537851	572	9·972478	77	9·565373	650	10·434627	49
12	538194	572	972431	78	565763	649	434237	48
13	538538	571	972385	78	566153	649	433847	47
14	538880	571	972338	78	566542	649	433458	46
15	539223	570	972291	78	566932	648	433068	45
16	539565	570	972245	78	567320	648	432680	44
17	539907	569	972198	78	567709	647	432291	43
18	540249	569	972151	78	568098	647	431902	42
19	540590	568	972105	78	568486	646	431514	41
20	540931	568	972058	78	568873	646	431127	40
21	9·541272	567	9·972011	78	9·569261	645	10·430739	39
22	541613	567	971964	78	569648	645	430352	38
23	541953	566	971917	78	570035	645	429965	37
24	542293	566	971870	78	570422	644	429578	36
25	542632	565	971823	78	570809	644	429191	35
26	542971	565	971776	78	571195	643	428805	34
27	543310	564	971729	79	571581	643	428419	33
28	543649	564	971682	79	571967	642	428033	32
29	543987	563	971635	79	572352	642	427648	31
30	544325	563	971588	79	572738	642	427262	30
31	9·544663	562	9·971540	79	9·573123	641	10·426877	29
32	545000	562	971493	79	573507	641	426493	28
33	545338	561	971446	79	573892	640	426108	27
34	545674	561	971398	79	574276	640	425724	26
35	546011	560	971351	79	574660	639	425340	25
36	546347	560	971303	79	575044	639	424956	24
37	546683	559	971256	79	575427	639	424573	23
38	547019	559	971208	79	575810	638	424190	22
39	547354	558	971161	79	576193	638	423807	21
40	547689	558	971113	79	576576	637	423424	20
41	9·548024	557	9·971066	80	9·576958	637	10·423041	19
42	548359	557	971018	80	577341	636	422659	18
43	548693	556	970970	80	577723	636	422277	17
44	549027	556	970922	80	578104	636	421896	16
45	549360	555	970874	80	578486	635	421514	15
46	549693	555	970827	80	578867	635	421133	14
47	550026	554	970779	80	579248	634	420752	13
48	550359	554	970731	80	579629	634	420371	12
49	550692	553	970683	80	580009	634	419991	11
50	551024	553	970635	80	580389	633	419611	10
51	9·551356	552	9·970586	80	9·580769	633	10·419231	9
52	551687	552	970538	80	581149	632	418851	8
53	552018	552	970490	80	581528	632	418472	7
54	552349	551	970442	80	581907	632	418093	6
55	552680	551	970394	80	582286	631	417714	5
56	553010	550	970345	81	582665	631	417335	4
57	553341	550	970297	81	583043	630	416957	3
58	553670	549	970249	81	583422	630	416578	2
59	554000	549	970200	81	583800	629	416200	1
60	554329	548	970152	81	584177	629	415823	0
	Cosine		Sine		Cotang.		Tang.	M.

69 Degrees.

M.	Sine	D.	Cosine	D.	Tang.	D.	Cotang.	
0	9·554329	548	9·970152	81	9·584177	699	10·415823	60
1	554658	548	970103	81	584555	629	415445	59
2	554987	547	960055	81	584932	628	415068	58
3	555315	547	970006	81	585309	628	414691	57
4	555643	546	900957	81	585686	627	414314	56
5	555971	546	969900	81	586062	627	413938	55
6	556299	545	909860	81	586439	627	413501	54
7	556626	545	969811	81	586815	626	413185	53
8	556953	544	969769	81	587190	626	412810	52
9	557280	544	909714	81	587566	625	412434	51
10	557606	543	969665	81	587941	625	412059	50
11	9·557932	543	9·969616	82	9·588316	625	10·411684	49
12	558258	543	960567	82	588691	624	411309	48
13	558583	542	969518	82	589066	624	410934	47
14	558909	542	969469	82	589440	623	410560	46
15	559234	541	969420	82	589814	623	410186	45
16	559558	541	969370	82	590168	623	409812	44
17	559883	540	969321	82	590562	622	409438	43
18	560207	540	969272	82	590935	622	409065	42
19	560531	539	900923	82	591308	622	408692	41
20	560855	539	969173	82	591681	621	408319	40
21	9·561178	538	9·969124	82	9·592054	621	10·407946	39
22	561501	538	969075	82	592426	620	407574	38
23	561824	537	969025	82	592798	620	407202	37
24	562146	537	968976	82	593170	619	406829	36
25	562468	536	968926	83	593542	619	406458	35
26	562790	536	968877	83	593914	618	406086	34
27	563112	536	968827	83	594285	618	405715	33
28	563433	535	968777	83	594656	618	405344	32
29	563755	535	968728	83	595027	617	404973	31
30	564075	534	968678	83	595398	617	404602	30
31	9·564396	534	9·968628	83	9·595768	617	10·404232	29
32	564716	533	968578	83	596138	616	403862	28
33	565036	533	968528	83	596508	616	403492	27
34	565356	532	968479	83	596878	616	403122	26
35	565676	532	968429	83	597247	615	402753	25
36	565995	531	968379	83	597616	615	402384	24
37	566314	531	968329	83	597985	615	402015	23
38	566632	531	968278	83	598354	614	401646	22
39	566951	530	968228	84	598722	614	401278	21
40	567269	530	968178	84	599091	613	400909	20
41	9·567587	529	9·968128	84	9·599459	613	10·400541	19
42	567004	529	968078	84	599627	613	400173	18
43	568222	528	968027	84	600194	612	399806	17
44	568539	528	967977	84	600562	612	399438	16
45	568856	528	967927	84	600929	611	399071	15
46	569172	527	967876	84	601296	611	398704	14
47	569488	527	967826	84	601662	611	398338	13
48	569804	526	967775	84	602029	610	397971	12
49	570120	526	967725	84	602395	610	397605	11
50	570435	525	967674	84	602761	610	397239	10
51	9·570751	525	9·967624	84	9·603127	609	10·396873	9
52	571066	524	967573	84	603493	609	396507	8
53	571380	524	967522	85	603858	609	396142	7
54	571695	523	967471	85	604223	608	395777	6
55	572009	523	967421	85	604588	608	395412	5
56	572323	523	967370	85	604953	607	395047	4
57	572636	522	967319	85	605317	607	394683	3
58	572950	522	967268	85	605682	607	394318	2
59	573263	521	967217	85	606046	606	393954	1
60	573575	521	967166	85	606410	606	393590	0
	Cosine		Sine		Cotang.		Tang.	M.

M.	Sine	D.	Cosine	D.	Tang.	D.	Cotang.	
0	9·573575	521	9·967166	85	9·606410	606	10·393590	60
1	573888	520	967115	85	606773	606	393227	59
2	574200	520	967064	85	607137	605	392863	58
3	574512	519	967013	85	607500	605	392500	57
4	574824	519	966961	85	607863	604	392137	56
5	575136	519	966910	85	608225	604	391775	55
6	575447	518	966859	85	608588	604	391412	54
7	575758	518	966808	85	608950	603	391050	53
8	576069	517	966756	86	609312	603	390688	52
9	576379	517	966705	86	609674	603	390326	51
10	576689	516	966653	86	610036	602	389964	50
11	9·576999	516	9·966602	86	9·610397	602	10·389603	49
12	577309	516	966550	86	610759	602	389241	48
13	577618	515	966499	86	611120	601	388880	47
14	577927	515	966447	86	611480	601	388520	46
15	578236	514	966395	86	611841	601	388159	45
16	578545	514	966344	86	612201	600	387799	44
17	578853	513	966292	86	612561	600	387439	43
18	579162	513	966240	86	612921	600	387079	42
19	579470	513	966188	86	613281	599	386719	41
20	579777	512	966136	86	613641	599	386359	40
21	9·580085	512	9·966085	87	9·614000	598	10·386000	39
22	580392	511	966033	87	614359	598	385641	38
23	580699	511	965981	87	614718	598	385282	37
24	581005	511	965928	87	615077	597	384923	36
25	581312	510	965876	87	615435	597	384565	35
26	581618	510	965824	87	615793	597	384207	34
27	581924	509	965772	87	616151	596	383849	33
28	582229	509	965720	87	616509	596	383491	32
29	582535	509	965668	87	616867	596	383133	31
30	582840	508	965615	87	617224	595	382776	30
31	9·583145	508	9·965563	87	9·617589	595	10·382418	29
32	583449	507	965511	87	617939	595	382061	28
33	583754	507	965458	87	618295	594	381705	27
34	584058	506	965406	87	618652	594	381348	26
35	584361	506	965353	88	619008	594	380992	25
36	584665	506	965301	88	619364	593	380636	24
37	584968	505	965248	88	619721	593	380279	23
38	585272	505	965195	88	620076	593	379924	22
39	585574	504	965143	88	620432	592	379568	21
40	585877	504	965090	88	620787	592	379213	20
41	9·586179	503	9·965037	88	9·621142	592	10·378858	19
42	586482	503	964984	88	621497	591	378503	18
43	586783	503	964931	88	621852	591	378148	17
44	587085	502	964879	88	622207	590	377793	16
45	587386	502	964826	88	622561	590	377439	15
46	587688	501	964773	88	622915	590	377085	14
47	587989	501	964719	88	623269	589	376731	13
48	588289	501	964666	89	623623	589	376377	12
49	588590	500	964613	89	623976	589	376024	11
50	588890	500	964560	89	624330	588	375670	10
51	9·589190	499	9·964507	89	9·624683	588	10·375317	9
52	589489	499	964454	89	625036	588	374964	8
53	589789	499	964400	89	625388	587	374612	7
54	590088	498	964347	89	625741	587	374259	6
55	590387	498	964294	89	626093	587	373907	5
56	590686	497	964240	89	626445	586	373555	4
57	590984	497	964187	89	626797	586	373203	3
58	591282	497	964133	89	627149	586	372851	2
59	591580	496	964080	89	627501	585	372499	1
60	591878	496	964026	89	627852	585	372148	0
	Cosine		Sine		Cotang.		Tang.	

67 Degrees.

M.	Sine	D.	Cosine	D.	Tang.	D.	Cotang.	
0	9·591878	496	9·964026	89	9·627852	585	10·372148	60
1	592176	495	963972	89	628203	585	371797	59
2	592473	495	963919	89	628554	585	371446	58
3	592770	495	963865	90	628905	584	371095	57
4	503067	494	963811	90	629255	584	370745	56
5	593363	494	963757	90	629606	583	370394	55
6	593659	493	963704	90	629956	583	370044	54
7	593955	493	963650	90	630306	583	369694	53
8	594251	493	963596	90	630656	583	369344	52
9	594547	492	963542	90	631005	582	368995	51
10	594842	492	963488	90	631355	582	368645	50
11	9·595137	491	0·963434	90	9·631704	582	10·368296	49
12	595432	491	963379	90	632053	581	367947	48
13	595727	491	963325	90	632401	581	367599	47
14	596021	490	963271	90	632750	581	367250	46
15	596315	490	963217	90	633098	580	366902	45
16	596609	489	963163	90	633447	580	366553	44
17	596903	489	963108	91	633795	580	366205	43
18	597196	489	963054	91	634143	579	365857	42
19	597490	488	962999	91	634490	579	365510	41
20	597783	488	962945	91	634838	579	365162	40
21	9·598075	487	9·962890	91	9·635185	578	10·364815	39
22	598368	487	962836	91	635532	578	364468	38
23	598660	487	962781	91	635879	578	364121	37
24	598952	486	962727	91	636226	577	363774	36
25	599244	486	962672	91	636572	577	363428	35
26	599536	485	962617	91	636919	577	363081	34
27	599827	485	962562	91	637265	577	362735	33
28	600118	485	962508	91	637611	576	362389	32
29	600409	484	962453	91	637956	576	362044	31
30	600700	484	962398	92	638302	576	361698	30
31	9·600990	484	9·962343	92	9·638647	575	10·361353	29
32	601280	483	962288	92	638992	575	361008	28
33	601570	483	962233	92	639337	575	360663	27
34	601860	482	962178	92	639682	574	360318	26
35	602150	482	962123	92	640027	574	359973	25
36	602439	482	962067	92	640371	574	359629	24
37	602728	481	962012	92	640716	573	359284	23
38	603017	481	961957	92	641060	573	358940	22
39	603305	481	961902	92	641404	573	358596	21
40	603594	480	961846	92	641747	572	358253	20
41	9·603882	480	9·961791	92	9·642091	572	10·357909	19
42	604170	479	961735	92	642434	572	357566	18
43	604457	479	961680	92	642777	572	357223	17
44	604745	479	961624	93	643120	571	356880	16
45	605032	478	961569	93	643463	571	356537	15
46	605319	478	961513	03	643806	571	356194	14
47	605606	478	961458	93	644148	570	355852	13
48	605892	477	961402	93	644490	570	355510	12
49	606179	477	961346	93	644832	570	355168	11
50	606465	476	961290	93	645174	569	354826	10
51	9·606751	476	9·961235	93	9·645516	569	10·354484	9
52	607036	476	961179	93	645857	569	354143	8
53	607322	475	961123	93	646199	569	353801	7
54	607607	475	961067	93	646540	568	353460	6
55	607892	474	961011	93	646881	568	353119	5
56	608177	474	960955	93	647222	568	352778	4
57	608461	474	960899	93	647562	567	352438	3
58	608745	473	960843	94	647903	567	352097	2
59	609029	473	960786	94	648243	567	351757	1
60	609313	473	960730	94	648583	566	351417	0
	Cosine		Sine		Cotang.		Tang.	M.

66 Degrees.

M.	Sine	D.	Cosine	D.	Tang.	D.	Cotang.	
0	9·609313	473	9·960730	94	9·648583	566	10·351417	60
1	609597	472	960674	94	648923	566	351077	59
2	609880	472	960618	94	649263	566	350737	58
3	610164	472	960561	94	649602	566	350398	57
4	610447	471	960505	94	649942	565	350058	56
5	610729	471	960448	94	650281	565	349719	55
6	611012	470	960392	94	650620	565	349380	54
7	611294	470	960335	94	650959	564	349041	53
8	611576	470	960279	94	651297	564	348703	52
9	611858	469	960222	94	651636	564	348364	51
10	612140	469	960165	94	651974	563	348026	50
11	0·612421	469	9·960109	95	9·652312	563	10·347688	49
12	612702	468	960052	95	652650	563	347350	48
13	612983	468	959995	95	652988	563	347012	47
14	613264	467	959938	95	653326	562	346674	46
15	613545	467	959882	95	653663	562	346337	45
16	613825	467	959825	95	654000	562	346000	44
17	614105	466	959768	95	654337	561	345663	43
18	614385	466	959711	95	654674	561	345326	42
19	614665	466	959654	95	655011	561	344989	41
20	614944	465	959596	95	655348	561	344652	40
21	9·615223	465	9·959539	95	9·655684	560	10·344316	39
22	615502	465	959482	95	656020	560	343980	38
23	615781	464	959425	95	656356	560	343644	37
24	616060	464	959368	95	656692	559	343308	36
25	616338	464	959310	96	657028	559	342972	35
26	616616	463	959253	96	657364	559	342636	34
27	616894	463	959195	96	657699	559	342301	33
28	617172	462	959138	96	658034	558	341966	32
29	617450	462	959081	96	658369	558	341631	31
30	617727	462	959023	96	658704	558	341296	30
31	9·618004	461	9·958965	96	9·659039	558	10·340961	29
32	618281	461	958908	96	659373	557	340627	28
33	618558	461	958850	96	659708	557	340292	27
34	618834	460	958792	96	660042	557	339958	26
35	619110	460	958734	96	660376	557	339624	25
36	619386	460	958677	96	660710	556	339290	24
37	619662	459	958619	96	661043	556	338957	23
38	619938	459	958561	96	661377	556	338623	22
39	620213	459	958503	97	661710	555	338290	21
40	620488	458	958445	97	662043	555	337957	20
41	9·620763	458	9·958387	97	9·662376	555	10·337624	19
42	621038	457	958329	97	662709	554	337291	18
43	621313	457	958271	97	663042	554	336958	17
44	621587	457	958213	97	663375	554	336625	16
45	621861	456	958154	97	663707	554	336293	15
46	622135	456	958096	97	664039	553	335961	14
47	622409	456	958038	97	664371	553	335629	13
48	622682	455	957979	97	664703	553	335297	12
49	622956	455	957921	97	665035	553	334965	11
50	623229	455	957863	97	665366	552	334634	10
51	9·623502	454	9·957804	97	9·665697	552	10·334303	9
52	623774	454	957746	98	666029	552	333971	8
53	624047	454	957687	98	666360	551	333640	7
54	624319	453	957628	98	666691	551	333309	6
55	624591	453	957570	98	667021	551	332979	5
56	624863	453	957511	98	667352	551	332648	4
57	625135	452	957452	98	667682	550	332318	3
58	625406	452	957393	98	668013	550	331987	2
59	625677	452	957335	98	668343	550	331657	1
60	625948	451	957276	98	668672	550	331328	0
	Cosine		Sine		Cotang.		Tang.	M.

65 Degrees.

M.	Sine	D.	Cosine	D.	Tang.	D.	Cotang.	
0	9·625948	451	9·957276	98	9·668673	550	10·331327	60
1	626219	451	957217	98	669002	549	330998	59
2	626490	451	957158	98	669332	549	330668	58
3	626760	450	957099	98	669661	549	330339	57
4	627030	450	957040	98	669991	548	330009	56
5	627300	450	956981	98	670320	548	329680	55
6	627570	449	956921	99	670649	548	329351	54
7	627840	449	956862	99	670977	548	329023	53
8	628109	449	956803	99	671306	547	328694	52
9	628378	448	956744	99	671634	547	328366	51
10	628647	448	956684	99	671963	547	328037	50
11	9·628916	447	9·956625	99	9·672291	547	10·327709	49
12	629185	447	956566	99	672619	546	327381	48
13	629453	447	956506	99	672947	546	327053	47
14	629721	446	956447	99	673274	546	326726	46
15	629989	446	956387	99	673602	546	326398	45
16	630257	446	956327	99	673929	545	326071	44
17	630524	446	956268	99	674257	545	325743	43
18	630792	445	956208	100	674584	545	325416	42
19	631059	445	956148	100	674910	544	325090	41
20	631326	445	956089	100	675237	544	324763	40
21	9·631593	444	9·956029	100	9·675564	544	10·324436	39
22	631859	444	955969	100	675890	544	324110	38
23	632125	444	955909	100	676216	543	323784	37
24	632392	443	955849	100	676543	543	323457	36
25	632658	443	955789	100	676869	543	323131	35
26	632923	443	955729	100	677194	543	322806	34
27	633189	442	955669	100	677520	542	322480	33
28	633454	442	955609	100	677846	542	322154	32
29	633719	442	955548	100	678171	542	321829	31
30	633984	441	955488	100	678496	542	321504	30
31	9·634249	441	9·955428	101	9·678821	541	10·321179	29
32	634514	440	955368	101	679146	541	320854	28
33	634778	440	955307	101	679471	541	320529	27
34	635042	440	955247	101	679795	541	320205	26
35	635306	439	955186	101	680120	540	319880	25
36	635570	439	955126	101	680444	540	319556	24
37	635834	439	955065	101	680768	540	319232	23
38	636097	438	955005	101	681092	540	318908	22
39	636360	438	954944	101	681416	539	318584	21
40	636623	438	954883	101	681740	539	318260	20
41	9·636886	437	9·954823	101	9·682063	539	10·317937	19
42	637148	437	954762	101	682387	539	317613	18
43	637411	437	954701	101	682710	538	317290	17
44	637673	437	954640	101	683033	538	316967	16
45	637935	436	954579	101	683356	538	316644	15
46	638197	436	954518	102	683679	538	316321	14
47	638458	436	954457	102	684001	537	315999	13
48	638720	435	954396	102	684324	537	315676	12
49	638981	435	954335	102	684646	537	315354	11
50	639242	435	954274	102	684968	537	315032	10
51	9·639503	434	9·954213	102	9·685290	536	10·314710	9
52	639764	434	954152	102	685612	536	314388	8
53	640024	434	954090	102	685934	536	314066	7
54	640284	433	954029	102	686255	536	313745	6
55	640544	433	953968	102	686577	535	313423	5
56	640804	433	953906	102	686898	535	313102	4
57	641064	432	953845	102	687219	535	312781	3
58	641324	432	953783	102	687540	535	312460	2
59	641584	432	953722	103	687861	534	312139	1
60	641842	431	953660	103	688182	534	311818	0
	Cosine		Sine		Cotang.		Tang.	M.

64 Degrees.

M.	Sine	D.	Cosine	D.	Tang.	D.	Cotang.	
0	9·641842	431	9·853660	103	9·688182	534	10·311818	60
1	642101	431	953509	103	688502	534	311498	59
2	642360	421	953537	103	688823	534	311177	58
3	642618	430	953475	103	689143	533	310857	57
4	642877	430	953413	103	689463	533	310537	56
5	643135	430	953352	103	689783	533	310217	55
6	643393	430	953290	103	690103	533	309897	54
7	643650	499	953229	103	690423	533	309577	53
8	643908	429	953166	103	690742	532	309258	52
9	644165	429	953104	103	691062	532	308938	51
10	644423	428	953042	103	691381	532	308619	50
11	9·644680	428	9·952980	104	9·691700	531	10·308300	49
12	644936	428	952918	104	692019	531	307981	48
13	645193	427	952855	104	692338	531	307662	47
14	645450	427	952793	104	692656	531	307344	46
15	645706	427	952731	104	692975	531	307025	45
16	645962	426	952669	104	693293	530	306707	44
17	646218	426	952606	104	693612	530	306388	43
18	646474	426	952544	104	693930	530	306070	42
19	646729	425	952481	104	694248	530	305752	41
20	646964	425	952419	104	694566	529	305434	40
21	9·647240	425	9·952356	104	9·694883	529	10·305117	39
22	647494	424	952294	104	695201	529	304790	38
23	647749	424	952231	104	695518	529	304482	37
24	648004	424	952168	105	695836	529	304164	36
25	648258	424	952106	105	696153	528	303847	35
26	648512	423	952043	105	696470	528	303530	34
27	648766	423	951980	105	696787	528	303213	33
28	649020	423	951917	105	697103	528	302897	32
29	649274	422	951854	105	697420	527	302580	31
30	649527	422	951791	105	697736	527	302264	30
31	9·649781	422	9·951728	105	9·698053	527	10·301947	29
32	650034	422	951665	105	698369	527	301631	28
33	650287	421	951602	105	698685	526	301315	27
34	650539	421	951539	105	699001	526	300999	26
35	650792	421	951476	105	699316	526	300684	25
36	651044	420	951412	105	699632	526	300368	24
37	651297	420	951349	106	699947	526	300053	23
38	651549	420	951286	106	700263	525	299737	22
39	651800	419	951222	106	700578	525	299422	21
40	652052	419	951159	106	700893	525	299107	20
41	9·652304	419	9·951096	106	9·701208	524	10·298702	19
42	652555	418	951032	106	701523	524	298477	18
43	652806	418	950968	106	701837	524	298163	17
44	653057	418	950905	106	702152	524	297848	16
45	653308	418	950841	106	702466	524	297534	15
46	653558	417	950778	106	702780	523	297220	14
47	653808	417	950714	106	703095	523	296905	13
48	654059	417	950650	106	703409	523	296591	12
49	654309	416	950586	106	703723	523	296277	11
50	654558	416	950522	107	704036	522	295964	10
51	9·654808	416	9·950458	107	9·704350	522	10·295650	9
52	655058	416	950394	107	704663	522	295337	8
53	655307	415	950330	107	704977	522	295023	7
54	655556	415	950266	107	705290	522	294710	6
55	655805	415	950202	107	705603	521	294397	5
56	656054	414	950138	107	705916	521	294084	4
57	656302	414	950074	107	706228	521	293772	3
58	656551	414	950010	107	706541	521	293459	2
59	656799	413	949945	107	706854	521	293146	1
60	657047	413	949881	107	707166	520	292834	0
	Cosine		Sine		Cotang.		Tang.	M.

63 Degrees.

M.	Sine	D.	Cosine	D.	Tang.	D.	Cotang.	
0	9·657047	413	9·949881	107	9·707166	520	10·292834	60
1	657295	413	949816	107	707478	520	292522	59
2	657542	412	949752	107	707790	520	292210	58
3	657790	412	949688	108	708102	520	291898	57
4	658037	412	949623	108	708414	519	291586	56
5	658284	412	949558	108	708726	519	291274	55
6	658531	411	949494	108	709037	519	290963	54
7	658778	411	949429	108	709349	519	290651	53
8	659025	411	949364	108	709660	519	290340	52
9	659271	410	949300	108	709971	518	290029	51
10	659517	410	949235	108	710282	518	289718	50
11	9·659763	410	9·949170	108	9·710593	518	10·289407	49
12	660009	409	949105	108	710904	518	289096	48
13	660255	409	949040	108	711215	518	288785	47
14	660501	409	948975	108	711525	517	288475	46
15	660746	409	948910	108	711836	517	288164	45
16	660991	408	948845	108	712146	517	287854	44
17	661236	408	948780	109	712456	517	287544	43
18	661481	408	948715	109	712766	516	287234	42
19	661726	407	948650	109	713076	516	286924	41
20	661970	407	948584	109	713386	516	286614	40
21	9·662214	407	9·948519	109	9·713696	516	10·286304	39
22	662459	407	948454	109	714005	516	285995	38
23	662703	406	948388	109	714314	515	285686	37
24	662946	406	948323	109	714624	515	285376	36
25	663190	406	948257	109	714933	515	285067	35
26	663433	405	948192	109	715242	515	284758	34
27	663677	405	948126	109	715551	514	284449	33
28	663920	405	948060	109	715860	514	284140	32
29	664162	405	947995	110	716168	514	283832	31
30	664406	404	947929	110	716477	514	283523	30
31	9·664648	404	9·947863	110	9·716785	514	10·283215	29
32	664891	404	947797	110	717093	513	282907	28
33	665133	403	947731	110	717401	513	282599	27
34	665375	403	947665	110	717709	513	282291	26
35	665617	403	947600	110	718017	513	281983	25
36	665859	402	947533	110	718325	513	281675	24
37	666100	402	947467	110	718633	512	281367	23
38	666342	402	947401	110	718940	512	281060	22
39	666583	402	947335	110	719248	512	280752	21
40	666824	401	947269	110	719555	512	280445	20
41	9·667065	401	9·947203	110	9·719862	512	10·280138	19
42	667305	401	947136	111	720169	511	279831	18
43	667546	401	947070	111	720476	511	279524	17
44	667786	400	947004	111	720783	511	279217	16
45	668027	400	946937	111	721089	511	278911	15
46	668267	400	946871	111	721396	511	278604	14
47	668506	399	946804	111	721702	510	278298	13
48	668746	399	946738	111	722009	510	277991	12
49	668986	399	946671	111	722315	510	277685	11
50	669225	399	946604	111	722621	510	277379	10
51	9·669464	398	9·946538	111	9·722927	510	10·277073	9
52	669703	398	946471	111	723232	509	276768	8
53	669942	398	946404	111	723538	509	276462	7
54	670181	397	946337	111	723844	509	276156	6
55	670419	397	946270	112	724149	509	275851	5
56	670658	397	946203	112	724454	509	275546	4
57	670896	397	946136	112	724759	508	275241	3
58	671134	396	946069	112	725065	508	274935	2
59	671372	396	946002	112	725369	508	274631	1
60	671609	396	945935	112	725674	508	274326	0
	Cosine		Sine		Cotang.		Tang.	M.

M.	Sine	D.	Cosine	D.	Tang.	D.	Cotang.	
0	9·671609	306	9·945935	112	9·725674	508	10·274326	60
1	671847	395	945868	112	725979	509	274021	59
2	672084	395	945800	112	726284	507	273716	58
3	672321	395	945733	112	726588	507	273412	57
4	672558	395	945666	112	726892	507	273108	56
5	672795	394	945598	112	727197	507	272803	55
6	673032	394	945531	112	727501	507	272499	54
7	673268	394	945464	113	727805	506	272195	53
8	673505	394	945396	113	728109	506	271891	52
9	673741	393	945328	113	728412	506	271588	51
10	673977	393	945261	113	728716	506	271284	50
11	9·674213	393	9·945193	113	9·729020	506	10·270980	49
12	674448	392	945125	113	729323	505	270677	48
13	674684	392	945058	113	729626	505	270374	47
14	674919	392	944990	113	729929	505	270071	46
15	675155	392	944922	113	730233	505	269767	45
16	675390	391	944854	113	730535	505	269465	44
17	675624	391	944786	113	730838	504	269162	43
18	675859	391	944718	113	731141	504	268859	42
19	676094	391	944650	113	731444	504	268556	41
20	676328	390	944582	114	731746	504	268254	40
21	9·676562	390	9·944514	114	9·732048	504	10·267952	39
22	676796	390	944446	114	732351	503	267649	38
23	677030	390	944377	114	732653	503	267347	37
24	677264	389	944309	114	732955	503	267045	36
25	677498	389	944241	114	733257	503	266743	35
26	677731	389	944172	114	733558	503	266442	34
27	677964	388	944104	114	733860	502	266140	33
28	678197	388	944036	114	734162	502	265838	32
29	678430	388	943967	114	734463	502	265537	31
30	678663	388	943899	114	734764	502	265236	30
31	9·678895	387	9·943830	114	9·735066	502	10·264934	29
32	679128	387	943761	114	735367	502	264633	28
33	679360	387	943693	115	735668	501	264332	27
34	679592	387	943624	115	735969	501	264031	26
35	679824	386	943555	115	736269	501	263731	25
36	680056	386	943486	115	736570	501	263430	24
37	680288	386	943417	115	736871	501	263129	23
38	680519	385	943348	115	737171	500	262829	22
39	680750	385	943279	115	737471	500	262529	21
40	680982	385	943210	115	737771	500	262229	20
41	9·681213	385	9·943141	115	9·738071	500	10·261929	19
42	681443	384	943072	115	738371	500	261629	18
43	681674	384	943003	115	738671	499	261329	17
44	681905	384	942934	115	738971	499	261029	16
45	682135	384	942864	115	739271	499	260729	15
46	682365	383	942795	116	739570	499	260430	14
47	682505	383	942726	116	739870	499	260130	13
48	682925	383	942656	116	740169	499	259831	12
49	683055	383	942587	116	740468	498	259532	11
50	683284	382	942517	116	740767	498	259233	10
51	9·683514	382	9·942448	116	9·741066	498	10·258934	9
52	683743	382	942378	116	741365	498	258635	8
53	683972	382	942308	116	741664	498	258336	7
54	684201	381	942239	116	741962	497	258038	6
55	684430	381	942169	116	742261	497	257739	5
56	684658	381	942099	116	742559	497	257441	4
57	684887	380	942029	116	742858	497	257142	3
58	685115	380	941959	116	743156	497	256844	2
59	685343	380	941889	117	743454	497	256546	1
60	685571	380	941819	117	743752	496	256248	0
	Cosine		Sine		Cotang.		Tang.	M.

61 Degrees.

M.	Sine	D.	Cosine	D.	Tang.	D.	Cotang.	
0	9·685571	380	9·941819	117	9·743759	496	10·256248	60
1	685799	379	941749	117	744050	496	255950	59
2	686027	379	941679	117	744348	496	255652	58
3	686254	379	941609	117	744645	496	255355	57
4	686482	379	941539	117	744943	496	255057	56
5	686709	378	941469	117	745240	496	254760	55
6	686936	378	941398	117	745538	495	254462	54
7	687163	378	941328	117	745835	495	254165	53
8	687389	378	941258	117	746132	495	253868	52
9	687616	377	941187	117	746429	495	253571	51
10	687843	377	941117	117	746726	495	253274	50
11	9·688069	377	9·941046	118	9·747023	494	10·252977	49
12	688295	377	940975	118	747319	494	252681	48
13	688521	376	940905	118	747616	494	252384	47
14	688747	376	940834	118	747913	494	252087	46
15	688972	376	940763	118	748209	494	251791	45
16	689198	376	940693	118	748505	493	251405	44
17	689423	375	940622	118	748801	493	251199	43
18	689648	375	940551	118	749097	493	250903	42
19	689873	375	940480	118	749393	493	250607	41
20	690098	375	940409	118	749689	493	250311	40
21	9·690323	374	9·940338	118	9·749985	493	10·250015	39
22	690548	374	940267	118	750281	492	249719	38
23	690772	374	940196	118	750576	492	249424	37
24	690996	374	940125	119	750872	492	249128	36
25	691220	373	940054	119	751167	492	248833	35
26	691444	373	939982	119	751462	492	248538	34
27	691668	373	939911	119	751757	492	248243	33
28	691899	373	939840	119	752052	491	247948	32
29	692115	372	939768	119	752347	491	247653	31
30	692339	372	939697	119	752642	491	247358	30
31	9·692562	372	9·939625	119	9·752937	491	10·247063	29
32	692785	371	939554	119	753231	491	246769	28
33	693008	371	939482	119	753526	491	246474	27
34	693231	371	939410	119	753820	490	246180	26
35	693453	371	939339	119	754115	490	245885	25
36	693676	370	939267	120	754409	490	245591	24
37	693898	370	939195	120	754703	490	245297	23
38	694120	370	939123	120	754997	490	245003	22
39	694342	370	939052	120	755291	490	244709	21
40	694564	369	938980	120	755585	489	244415	20
41	9·694786	369	9·938908	120	9·755878	489	10·244122	19
42	695007	369	938836	120	756172	489	243828	18
43	695229	369	938763	120	756465	489	243535	17
44	695450	368	938691	120	756759	489	243241	16
45	695671	368	938619	120	757052	489	242948	15
46	695893	368	938547	120	757345	488	242655	14
47	696113	368	938475	120	757638	488	242362	13
48	696334	367	938402	121	757931	488	242069	12
49	696554	367	938330	121	758224	488	241776	11
50	696775	367	938258	121	758517	488	241483	10
51	9·696995	367	9·938185	121	9·758810	488	10·241190	9
52	697215	366	938113	121	759102	487	240898	8
53	697435	366	938040	121	759395	487	240605	7
54	697654	366	937967	121	759687	487	240313	6
55	697874	366	937895	121	759979	487	240021	5
56	698094	365	937822	121	760272	487	239728	4
57	698313	365	937749	121	760564	487	239436	3
58	698532	365	937676	121	760856	486	239144	2
59	698751	365	937604	121	761148	486	238852	1
60	698970	364	937531	121	761439	486	238561	0
	Cosine		Sine		Cotang.		Tang.	M.

60 Degrees.

M.	Sine	D.	Cosine	D.	Tang.	D.	Cotang.	
0	9·698970	364	9·937531	121	9·761439	486	10·238561	60
1	699189	364	937458	122	761731	486	238269	59
2	699407	364	937385	122	762093	486	237977	58
3	699626	364	937312	122	762314	486	237686	57
4	699844	363	937238	122	762606	485	237394	56
5	700062	363	937165	122	762897	485	237103	55
6	700280	363	937092	122	763188	485	236812	54
7	700498	363	937019	122	763479	485	236521	53
8	700716	363	936946	122	763770	485	236230	52
9	700933	362	936872	122	764061	485	235939	51
10	701151	362	936799	122	764352	484	235648	50
11	9·701368	362	9·936725	122	9·764643	484	10·235357	49
12	701585	362	936652	123	764933	484	235067	48
13	701802	361	936578	123	765224	484	234776	47
14	702019	361	936505	123	765514	484	234486	46
15	702236	361	936431	123	765805	484	234195	45
16	702452	361	936357	123	766095	484	233905	44
17	702669	360	936284	123	766385	483	233615	43
18	702885	360	936210	123	766675	483	233325	42
19	703101	360	936136	123	766965	483	233035	41
20	703317	360	936062	123	767255	483	232745	40
21	9·703533	359	9·935988	123	9·767545	483	10·232455	39
22	703749	359	935914	123	767834	483	232166	38
23	703964	359	935840	123	768124	482	231876	37
24	704179	359	935766	124	768413	482	231587	36
25	704395	359	935692	124	768703	482	231297	35
26	704610	358	935618	124	768992	482	231008	34
27	704825	358	935543	124	769281	482	230719	33
28	705040	358	935469	124	769570	482	230430	32
29	705254	358	935395	124	769860	481	230140	31
30	705469	357	935320	124	770148	481	229852	30
31	9·705683	357	9·935246	124	9·770437	481	10·229563	29
32	705898	357	935171	124	770726	481	229274	28
33	706112	357	935097	124	771015	481	228985	27
34	706326	356	935022	124	771303	481	228697	26
35	706539	356	934948	124	771592	481	228408	25
36	706753	356	934873	124	771880	480	228120	24
37	706967	356	934798	125	772168	480	227832	23
38	707180	355	934723	125	772457	480	227543	22
39	707393	355	934649	125	772745	480	227255	21
40	707606	355	934574	125	773033	480	226967	20
41	9·707819	355	9·934499	125	9·773321	480	10·226679	19
42	708032	354	934424	125	773608	479	226392	18
43	708245	354	934349	125	773896	479	226104	17
44	708458	354	934274	125	774184	479	225816	16
45	708670	354	934199	125	774471	479	225529	15
46	708882	353	934123	125	774759	479	225241	14
47	709094	353	934048	125	775046	479	224954	13
48	709306	353	933973	125	775333	479	224667	12
49	709518	353	933898	126	775621	478	224379	11
50	709730	353	933822	126	775908	478	224092	10
51	9·709941	352	9·933747	126	9·776195	478	10·223805	9
52	710153	352	933671	126	776482	478	223518	8
53	710364	352	933596	126	776769	478	223231	7
54	710575	352	933520	126	777055	478	222945	6
55	710786	351	933445	126	777342	478	222658	5
56	710997	351	933369	126	777628	477	222372	4
57	711208	351	933293	126	777915	477	222085	3
58	711419	351	933217	126	778201	477	221799	2
59	711629	350	933141	126	778487	477	221512	1
60	711839	350	933066	126	778774	477	221226	0
	Cosine		Sine		Cotang.		Tang.	M.

59 Degrees.

M.	Sine	D.	Cosine	D.	Tang.	D.	Cotang.	
0	9·711839	350	9·933066	126	9·778774	477	10·221226	60
1	712050	350	932940	127	779060	477	220940	59
2	712260	350	932914	127	779346	476	220654	58
3	712469	349	932838	127	779632	476	220368	57
4	712679	349	932762	127	779918	476	220082	56
5	712889	349	932685	127	780203	476	219797	55
6	713098	349	932609	127	780489	476	219511	54
7	713308	349	932533	127	780775	476	219225	53
8	713517	348	932457	127	781060	476	218940	52
9	713726	348	932380	127	781346	475	218654	51
10	713935	348	932304	127	781631	475	218369	50
11	9·714144	348	9·932228	127	9·781916	475	10·218084	49
12	714352	347	932151	127	782201	475	217799	48
13	714561	347	932075	128	782486	475	217514	47
14	714769	347	931998	128	782771	475	217229	46
15	714978	347	931921	128	783056	475	216944	45
16	715186	347	931845	128	783341	475	216659	44
17	715394	346	931768	128	783626	474	216374	43
18	715602	346	931691	128	783910	474	216090	42
19	715809	346	931614	128	784195	474	215805	41
20	716017	346	931537	128	784479	474	215521	40
21	9·716224	345	9·931460	128	9·784764	474	10·215236	39
22	716432	345	931383	128	785048	474	214952	38
23	716639	345	931306	128	785332	473	214668	37
24	716846	345	931229	129	785616	473	214384	36
25	717053	345	931152	129	785900	473	214100	35
26	717259	344	931075	129	786184	473	213816	34
27	717466	344	930998	129	786468	473	213532	33
28	717673	344	930921	129	786752	473	213248	32
29	717879	344	930843	129	787036	473	212964	31
30	718085	343	930766	129	787319	472	212681	30
31	9·718291	343	9·930688	129	9·787603	472	10·212397	29
32	718497	343	930611	129	787880	472	212114	28
33	718703	343	930533	129	788170	472	211830	27
34	718909	343	930456	129	788453	472	211547	26
35	719114	342	930378	129	788736	472	211264	25
36	719320	342	930300	130	789019	472	210981	24
37	719525	342	930223	130	789302	471	210698	23
38	719730	342	930145	130	789585	471	210415	22
39	719935	341	930067	130	789868	471	210132	21
40	720140	341	929989	130	790151	471	209849	20
41	9·720345	341	9·929911	130	9·790433	471	10·209567	19
42	720549	341	929833	130	790716	471	209284	18
43	720754	340	929755	130	790999	471	209001	17
44	720958	340	929677	130	791281	471	208719	16
45	721162	340	929599	130	791563	470	208437	15
46	721366	340	929521	130	791846	470	208154	14
47	721570	340	929442	130	792128	470	207872	13
48	721774	339	929364	131	792410	470	207590	12
49	721978	339	929286	131	792692	470	207308	11
50	722181	339	929207	131	792974	470	207026	10
51	9·722385	339	9·929129	131	9·793256	470	10·206744	9
52	722588	339	929050	131	793538	469	206462	8
53	722791	338	928972	131	793819	469	206181	7
54	722994	338	928893	131	794101	469	205899	6
55	723197	338	928815	131	794383	469	205617	5
56	723400	338	928736	131	794664	469	205336	4
57	723603	337	928657	131	794945	469	205055	3
58	723805	337	928578	131	795227	469	204773	2
59	724007	337	928499	131	795508	468	204492	1
60	724210	337	928420	131	795789	468	204211	0
	Cosine		Sine		Cotang.		Tang.	M.

M.	Sine	D.	Cosine	D.	Tang.	D.	Cotang.	
0	9·724210	337	9·928420	132	9·795789	468	10·204211	60
1	724412	337	928342	132	796070	468	203930	59
2	724614	336	928263	132	796351	468	203649	58
3	724816	336	928183	132	796632	468	203368	57
4	725017	336	928104	132	796913	468	203087	56
5	725219	336	928025	132	797194	468	202806	55
6	725420	335	927946	132	797475	468	202525	54
7	725622	335	927867	132	797755	468	202245	53
8	725823	335	927787	132	798036	467	201964	52
9	726024	335	927708	132	798316	467	201684	51
10	726225	335	927629	132	798596	467	201404	50
11	9·726426	334	9·927549	132	9·798877	467	10·201123	49
12	726626	334	927470	133	799157	467	200843	48
13	726827	334	927390	133	799437	467	200563	47
14	727027	334	927310	133	799717	467	200283	46
15	727228	334	·927231	133	799997	466	200003	45
16	727428	333	927151	133	800277	466	199723	44
17	727628	333	927071	133	800557	466	199443	43
18	727828	333	926991	133	800836	466	199164	42
19	728027	333	926911	133	801116	466	198884	41
20	728227	333	926831	133	801396	466	198604	40
21	9·728427	332	9·926751	133	9·801675	466	10·198325	39
22	728626	332	926671	133	801955	466	198045	38
23	728825	332	926591	133	802234	465	197766	37
24	729024	332	926511	134	802513	465	197487	36
25	729223	331	926431	134	802792	465	197208	35
26	729422	331	926351	134	803072	465	196928	34
27	729621	331	926270	134	803351	465	196649	33
28	729820	331	926190	134	803630	465	196370	32
29	730018	330	926110	134	803908	465	196092	31
30	730216	330	926029	134	804187	465	195813	30
31	9 730415	330	9·925949	134	9·804466	464	10·195534	29
32	730613	330	925868	134	804745	464	195255	28
33	730811	330	925788	134	805023	464	194977	27
34	731009	329	925707	134	805302	464	194698	26
35	731206	329	925626	134	805580	464	194420	25
36	731404	329	925545	135	805859	464	194141	24
37	731602	329	925465	135	806137	464	193863	23
38	731799	329	925384	135	806415	463	193585	22
39	731996	328	925303	135	806693	463	193307	21
40	732193	328	925222	135	806971	463	193029	20
41	9 732390	328	9·925141	135	9·807249	463	10·192751	19
42	732587	328	925060	135	807527	463	192473	18
43	732784	328	924979	135	807805	463	192195	17
44	732980	327	924897	135	808083	463	191917	16
45	733177	327	924816	135	808361	463	191639	15
46	733373	327	924735	136	808638	462	191362	14
47	733569	327	924654	136	808916	462	191084	13
48	733765	327	924572	136	809193	462	190807	12
49	733961	326	924491	136	809471	462	190529	11
50	734157	326	924409	136	809748	462	190252	10
51	9 734353	326	9·924328	136	9·810025	462	10·189975	9
52	734549	326	924246	136	810302	462	189698	8
53	734744	325	924164	136	810580	462	189420	7
54	734939	325	924083	136	810857	462	189143	6
55	735135	325	924001	136	811134	461	188866	5
56	735330	325	923919	136	811410	461	188590	4
57	735525	325	923837	136	811687	461	188313	3
58	735719	324	923755	137	811964	461	188036	2
59	735014	324	923673	137	812241·	461	187759	1
60	736109	324	923591	137	812517	461	187483	0
	Cosine		Sine		Cotang.		Tang.	M.

57 Degrees.

M.	Sine	D.	Cosine	D.	Tang.	D.	Cotang.	
0	9·736109	324	9·923591	137	9·812517	461	10·187482	60
1	736303	324	923509	137	812794	461	187206	59
2	736498	324	923427	137	813070	461	186930	58
3	736692	323	923345	137	813347	460	186653	57
4	736886	323	923263	137	813623	460	186377	56
5	737080	323	923181	137	813899	460	186101	55
6	737274	323	923098	137	814175	460	185825	54
7	737467	323	923016	137	814452	460	185548	53
8	737661	322	922933	137	814728	460	185272	52
9	737855	322	922851	137	815004	460	184996	51
10	738048	322	922708	138	815279	460	184721	50
11	9·738241	322	9·922686	138	9·815555	459	10·184445	49
12	738434	322	922603	138	815831	459	184169	48
13	738627	321	922520	138	816107	459	183893	47
14	738820	321	922438	138	816382	459	183618	46
15	739013	321	922355	138	816658	459	183342	45
16	739206	321	922272	138	816933	459	183067	44
17	739398	321	922189	138	817209	459	182791	43
18	739590	320	922106	138	817484	459	182516	42
19	739783	320	922023	138	817759	459	182241	41
20	739975	320	921040	138	818035	458	181965	40
21	9·740167	320	9·921857	139	9·818310	458	10·181690	39
22	740359	320	921774	139	818585	458	181415	38
23	740550	319	921691	139	818860	458	181140	37
24	740742	319	921607	139	819135	458	180865	36
25	740934	319	921524	139	819410	458	180590	35
26	741125	319	921441	139	819684	458	180316	34
27	741316	319	921357	139	819959	458	180041	33
28	741508	318	921274	139	820234	458	179766	32
29	741699	318	921190	139	820508	457	179492	31
30	741889	318	921107	139	820783	457	179217	30
31	9·742080	318	9·921023	139	9·821057	457	10·178943	29
32	742271	318	920939	140	821332	457	178668	28
33	742462	317	920856	140	821606	457	178394	27
34	742652	317	920772	140	821880	457	178120	26
35	742842	317	920688	140	822154	457	177846	25
36	743033	317	920604	140	822429	457	177571	24
37	743223	317	920520	140	822703	457	177297	23
38	743413	316	920436	140	822977	456	177023	22
39	743602	316	920352	140	823250	456	176750	21
40	743792	316	920268	140	823524	456	176476	20
41	9·743982	316	9·920184	140	9·823798	456	10·176202	19
42	744171	316	920099	140	824072	456	175928	18
43	744361	315	920015	140	824345	456	175655	17
44	744550	315	919931	141	824619	456	175381	16
45	744739	315	919846	141	824893	456	175107	15
46	744928	315	919762	141	825166	456	174834	14
47	745117	315	919677	141	825439	455	174561	13
48	745306	314	919593	141	825713	455	174287	12
49	745494	314	919508	141	825986	455	174014	11
50	745683	314	919424	141	826259	455	173741	10
51	9·745871	314	9·919339	141	9·826532	455	10·173468	9
52	746059	314	919254	141	826805	455	173195	8
53	746248	313	919169	141	827078	455	172922	7
54	746436	313	919085	141	827351	455	172649	6
55	746624	313	919000	141	827624	455	172376	5
56	746812	313	918915	142	827897	454	172103	4
57	746999	313	918830	142	828170	454	171830	3
58	747187	312	918745	142	828442	454	171558	2
59	747374	312	918659	142	828715	454	171285	1
60	747562	312	918574	142	828987	454	171013	0
	Cosine		Sine		Cotang.		Tang.	M.

56 Degrees.

M.	Sine	D.	Cosine	D.	Tang.	D.	Cotang.	
0	9·747562	312	9·918574	142	9·828987	454	10·171013	60
1	747749	312	918489	142	829260	454	170740	59
2	747936	312	918404	142	829532	454	170468	58
3	748123	311	918318	142	829805	454	170195	57
4	748310	311	918233	142	830077	454	169923	56
5	748497	311	918147	142	830349	453	169651	55
6	748683	311	918062	142	830621	453	169379	54
7	748870	311	917976	143	830893	453	169107	53
8	749056	310	917891	143	831165	453	168835	52
9	749243	310	917805	143	831437	453	168563	51
10	749429	310	917719	143	831709	453	168291	50
11	9·749615	310	9·917634	143	9·831981	453	10·168019	49
12	749801	310	917548	143	832253	453	167747	48
13	749987	309	917462	143	832525	453	167475	47
14	750172	309	917376	143	832796	453	167204	46
15	750358	309	917290	143	833068	452	166932	45
16	750543	309	917204	143	833339	452	166661	44
17	750729	309	917118	144	833611	452	166389	43
18	750914	308	917032	144	833882	452	166118	42
19	751099	308	916946	144	834154	452	165846	41
20	751284	308	916859	144	834425	452	165575	40
21	9·751469	308	9·916773	144	9·834696	452	10·165304	39
22	751654	308	916687	144	834967	452	165033	38
23	751839	308	916600	144	835238	452	164762	37
24	752023	307	916514	144	835509	452	164491	36
25	752208	307	916427	144	835780	451	164220	35
26	752392	307	916341	144	836051	451	163949	34
27	752576	307	916254	144	836322	451	163678	33
28	752760	307	916167	145	836593	451	163407	32
29	752944	306	916081	145	836864	451	163136	31
30	753128	306	915994	145	837134	451	162866	30
31	9·753312	306	9·915907	145	9·837405	451	10·162595	29
32	753495	306	915820	145	837675	451	162325	28
33	753679	306	915733	145	837946	451	162054	27
34	753862	305	915646	145	838216	451	161784	26
35	754046	305	915559	145	838487	450	161513	25
36	754229	305	915472	145	838757	450	161243	24
37	754412	305	915385	145	839027	450	160973	23
38	754595	305	915297	145	839297	450	160703	22
39	754778	304	915210	145	839568	450	·160432	21
40	754960	304	915123	146	839838	450	160162	20
41	9·755143	304	9·915035	146	9·840108	450	10·159892	19
42	755326	304	914948	146	840378	450	159622	18
43	755508	304	914860	146	840647	450	159353	17
44	755690	304	914773	146	840917	449	159083	16
45	755872	303	914685	146	841187	449	158813	15
46	756054	303	914598	146	841457	449	158543	14
47	756236	303	914510	146	841726	449	158274	13
48	756418	303	914422	146	841996	449	158004	12
49	756600	303	914334	146	842266	449	157734	11
50	756782	302	914246	147	842535	449	157465	10
51	9·756963	302	9·914158	147	9·842805	449	10·157195	9
52	757144	302	914070	147	843074	449	156926	8
53	757326	302	913982	147	843343	449	156657	7
54	757507	302	913894	147	843612	448	156388	6
55	757688	301	913806	147	843882	448	156118	5
56	757869	301	913718	147	844151	448	155849	4
57	758050	301	913630	147	844420	448	155580	3
58	758230	301	913541	147	844689	448	155311	2
59	758411	301	913453	147	844958	448	155042	1
60	758591	301	913365	147	845227	448	154773	0
	Cosine		Sine		Cotang.		Tang.	M.

55 Degrees.

M.	Sine	D.	Cosine	D.	Tang.	D.	Cotang.	
0	9·758591	301	9·913365	147	9·845237	448	10·154773	60
1	758772	300	913276	147	845496	448	154504	59
2	758952	300	913187	148	845764	448	154236	58
3	759133	300	913099	148	846033	448	153967	57
4	759313	300	913010	148	846302	448	153698	56
5	759492	300	912922	148	846570	447	153430	55
6	759672	299	912833	148	846839	447	153161	54
7	759852	299	912744	148	847107	447	152893	53
8	760031	299	912655	148	847376	447	152624	52
9	760211	299	912566	148	847644	447	152356	51
10	760390	299	912477	148	847913	447	152087	50
11	9·760569	298	9·912388	148	9·848181	447	10·151819	49
12	760748	298	912299	149	848449	447	151551	48
13	760927	298	912210	149	848717	447	151283	47
14	761106	298	912121	149	848986	447	151014	46
15	761285	298	912031	149	849254	447	150746	45
16	761464	298	911942	149	849522	447	150478	44
17	761642	297	911853	149	849790	446	150210	43
18	761821	297	911703	149	850058	·446	149942	42
19	761999	297	911674	149	850325	446	149675	41
20	762177	297	911584	149	850593	446	149407	40
21	9·762356	297	9·911495	149	9·850861	446	10·149139	39
22	762534	296	911405	149	851129	446	148871	38
23	762712	296	911315	150	851396	446	148604	37
24	762889	296	911226	150	851664	446	148336	36
25	763067	296	911136	150	851931	446	148069	35
26	763245	296	911046	150	852199	446	147801	34
27	763422	296	910956	150	852466	446	147534	33
28	763600	295	910866	150	852733	445	147267	32
29	763777	295	910776	150	853001	445	146999	31
30	763954	295	910686	150	853268	445	146732	30
31	9·764131	295	9·910596	150	9·853535	445	10·146465	29
32	764308	295	910506	150	853802	445	146198	28
33	764485	294	910415	150	854069	445	145931	27
34	764662	294	910325	151	854336	445	145664	26
35	764838	294	910235	151	854603	445	145397	25
36	765015	294	910144	151	854870	445	145130	24
37	765191	294	910054	151	855137	445	144863	23
38	765367	294	909963	151	855404	445	144596	22
39	765544	293	909873	151	855671	444	144329	21
40	765720	293	909782	151	855938	444	144062	20
41	9·765896	293	9·909691	151	9·856204	444	10·143796	19
42	766072	293	909601	151	856471	444	143529	18
43	766247	293	909510	151	856737	444	143263	17
44	766423	293	909419	151	857004	444	142996	16
45	766598	292	909328	152	857270	444	142730	15
46	766774	292	909237	152	857537	444	142463	14
47	766949	292	909146	152	857803	444	142197	13
48	767124	292	909055	152	858069	444	141931	12
49	767300	292	908964	152	858336	444	141664	11
50	767475	291	908873	152	858602	443	141398	10
51	9·767649	291	9·908781	152	9·858868	443	10·141112	9
52	767824	291	908690	152	859134	443	140866	8
53	767999	291	908599	152	859400	443	140600	7
54	768173	291	908507	152	859666	443	140334	6
55	768348	290	908416	153	859932	443	140068	5
56	768522	290	908324	153	860198	443	139802	4
57	768697	290	908233	153	860464	443	139536	3
58	768871	290	908141	153	860730	443	139270	2
59	769045	290	908049	153	860995	443	139005	1
60	769219	290	907958	153	861261	443	138739	0
	Cosine		Sine		Cotang.		Tang.	M.

54 Degrees.

M.	Sine	D.	Cosine	D.	Tang.	D.	Cotang.	
0	9·769219	290	9·907958	153	9·861261	443	10·138739	60
1	769393	289	907866	153	861527	443	138473	59
2	769566	289	907774	153	861792	442	138208	58
3	769740	289	907682	153	862058	442	137942	57
4	769913	289	907590	153	862323	442	137677	56
5	770087	289	907498	153	862589	442	137411	55
6	770260	288	907406	153	862854	442	137146	54
7	770433	288	907314	154	863119	442	136881	53
8	770606	288	907222	154	863385	442	136615	52
9	770779	288	907129	154	863650	442	136350	51
10	770952	288	907037	154	863915	442	136085	50
11	9·771125	288	9·906945	154	9·864180	442	10·135820	49
12	771298	287	906852	154	864445	442	135555	48
13	771470	287	906760	154	864710	442	135290	47
14	771643	287	906667	154	864975	441	135025	46
15	771815	287	906575	154	865240	441	134760	45
16	771987	287	906482	154	865505	441	134495	44
17	772159	287	906389	155	865770	441	134230	43
18	772331	286	906296	155	866035	441	133965	42
19	772503	286	906204	155	866300	441	133700	41
20	772675	286	906111	155	866564	441	133436	40
21	9·772847	286	9·906018	155	9·866829	441	10·133171	39
22	773018	286	905925	155	867094	441	132906	38
23	773190	286	905832	155	867358	441	132642	37
24	773361	285	905739	155	867623	441	132377	36
25	773533	285	905645	155	867887	441	132113	35
26	773704	285	905552	155	868152	440	131848	34
27	773875	285	905459	155	868416	440	131584	33
28	774046	285	905366	156	868680	440	131320	32
29	774217	285	905272	156	868945	440	131055	31
30	774388	284	905179	156	869209	440	130791	30
31	9·774558	284	9·905085	156	9·869473	440	10·130527	29
32	774729	284	904992	156	869737	440	130263	28
33	774899	284	904898	156	870001	440	129999	27
34	775070	284	904804	156	870265	440	129735	26
35	775240	284	904711	156	870529	440	129471	25
36	775410	283	904617	156	870793	440	129207	24
37	775580	283	904523	156	871057	440	128943	23
38	775750	283	904429	157	871321	440	128679	22
39	775920	283	904335	157	871585	439	128415	21
40	776090	283	904241	157	871849	439	128151	20
41	9·776250	283	9·904147	157	9·872112	439	10·127888	19
42	776429	282	904053	157	872376	439	127624	18
43	776598	282	903959	157	872640	439	127360	17
44	776768	282	903864	157	872903	439	127097	16
45	776937	282	903770	157	873167	439	126833	15
46	777106	282	903676	157	873430	439	126570	14
47	777275	281	903581	157	873694	439	126306	13
48	777444	281	903487	157	873957	439	126043	12
49	777613	281	903392	158	874220	439	125780	11
50	777781	281	903298	158	874484	439	125516	10
51	9·777950	281	9·903203	158	9·874747	439	10·125253	9
52	778119	281	903108	158	875010	439	124990	8
53	778287	280	903014	158	875273	438	124727	7
54	778455	280	902919	158	875536	438	124464	6
55	778624	280	902824	158	875800	438	124200	5
56	778792	280	902729	158	876063	438	123937	4
57	778960	280	902634	158	876326	438	123674	3
58	779128	280	902539	159	876589	438	123411	2
59	779295	279	902444	159	876851	438	123149	1
60	779463	279	902340	159	877114	438	122886	0
	Cosine		Sine		Cotang.		Tang.	M.

53 Degrees.

M.	Sine	D.	Cosine	D.	Tang.	D.	Cotang.	
0	9·779463	279	9·902349	159	9·877114	438	10·122886	60
1	779631	279	902253	159	877377	438	122623	59
2	779798	279	902158	159	877640	438	122360	58
3	779966	279	902063	159	877903	438	122097	57
4	780133	279	901967	159	878165	438	121835	56
5	780300	278	901872	159	878428	438	121572	55
6	780467	278	901776	159	878691	438	121309	54
7	780634	278	901681	159	878953	437	121047	53
8	780801	278	901585	159	879216	437	120784	52
9	780968	278	901490	159	879478	437	120522	51
10	781134	278	901394	160	879741	437	120259	50
11	9·781301	277	9·901298	160	9·880003	437	10·119997	49
12	781468	277	901202	160	880265	437	119735	48
13	781634	277	901106	160	880528	437	119472	47
14	781800	277	901010	160	880790	437	119210	46
15	781966	277	900914	160	881052	437	118948	45
16	782132	277	900818	160	881314	437	118686	44
17	782298	276	900722	160	881576	437	118424	43
18	782464	276	900626	160	881839	437	118161	42
19	782630	276	900529	160	882101	437	117899	41
20	782796	276	900433	161	882363	436	117637	40
21	9·782961	276	9·900337	161	9·882625	436	10·117375	39
22	783127	276	900240	161	882887	436	117113	38
23	783292	275	900144	161	883148	436	116852	37
24	783458	275	900047	161	883410	436	116590	36
25	783623	275	899951	161	883672	436	116328	35
26	783788	275	899854	161	883934	436	116066	34
27	783953	275	899757	161	884196	436	115804	33
28	784118	275	899660	161	884457	436	115543	32
29	784282	274	899564	161	884719	436	115281	31
30	784447	274	899467	162	884980	436	115020	30
31	9·784612	274	9·899370	162	9·885242	436	10·114758	29
32	784776	274	899273	162	885503	436	114407	28
33	784941	274	899176	162	885765	436	114235	27
34	785105	274	899078	162	886026	436	113974	26
35	785269	273	898981	162	886288	436	113712	25
36	785433	273	898884	162	886549	435	113451	24
37	785597	273	898787	162	886810	435	113190	23
38	785761	273	898689	162	887072	435	112928	22
39	785925	273	898592	162	887333	435	112667	21
40	786089	273	898494	163	887594	435	112406	20
41	9·786252	272	9·898397	163	9·887855	435	10·112145	19
42	786416	272	898299	163	888116	435	111884	18
43	786579	272	898202	163	888377	435	111623	17
44	786742	272	898104	163	888639	435	111361	16
45	786906	272	898006	163	888900	435	111100	15
46	787069	272	897908	163	889160	435	110840	14
47	787232	271	897810	163	889421	435	110579	13
48	787395	271	897712	163	889682	435	110318	12
49	787557	271	897614	163	889943	435	110057	11
50	787720	271	897516	163	890204	434	109796	10
51	9·787883	271	9·897418	164	9·890465	434	10·109535	9
52	788045	271	897320	164	890725	434	109275	8
53	788208	271	897222	164	890986	434	109014	7
54	788370	270	897123	164	891247	434	108753	6
55	788532	270	897025	164	891507	434	108493	5
56	788694	270	896926	164	891768	434	108232	4
57	788856	270	896828	164	892028	434	107972	3
58	789018	270	896729	164	892289	434	107711	2
59	789180	270	896631	164	892549	434	107451	1
60	789342	269	896532	164	892810	434	107190	0
	Cosine		Sine		Cotang.		Tang.	M.

52 Degrees.

M.	Sine	D.	Cosine	D.	Tang.	D.	Cotang.	
0	9·789342	289	9·896532	164	9·892810	434	10·107190	60
1	789504	289	896433	165	893070	434	106930	59
2	789665	290	896335	165	893331	434	106669	58
3	789827	290	896236	165	893591	434	106409	57
4	789988	290	896137	165	893851	434	106149	56
5	790149	289	896038	165	894111	434	105889	55
6	790310	288	895939	165	894371	434	105629	54
7	790471	288	895840	165	894632	433	105368	53
8	790632	288	895741	165	894892	433	105108	52
9	790793	288	895641	165	895152	433	104848	51
10	790954	288	895542	165	895412	433	104588	50
11	9·791115	288	9·895443	166	9·895672	433	10·104328	49
12	791275	287	895343	166	895932	433	104068	48
13	791436	267	895244	166	896192	433	103808	47
14	791596	267	895145	166	896452	433	103548	46
15	791757	267	895045	166	896712	433	103288	45
16	791917	267	894945	166	896971	433	103029	44
17	792077	267	894846	166	897231	433	102769	43
18	792237	266	894746	166	897491	433	102509	42
19	792397	266	894646	166	897751	433	102249	41
20	792557	266	894540	166	898010	433	101990	40
21	9·792716	266	9·894446	167	9·898270	433	10·101730	39
22	792876	266	894346	167	898530	433	101470	38
23	793035	266	894246	167	898789	433	101211	37
24	793195	265	894146	167	899049	432	100951	36
25	793354	265	894046	167	899308	432	100692	35
26	793514	265	893946	167	899568	432	100432	34
27	793673	265	893846	167	899827	432	100173	33
28	793832	265	893745	167	900086	432	099914	32
29	793991	265	893645	167	900346	432	099654	31
30	794150	264	893544	167	900605	432	099395	30
31	9·794308	264	9·893444	168	9·900864	432	10·099136	29
32	794467	264	893343	168	901124	432	098876	28
33	794626	264	893243	168	901383	432	098617	27
34	794784	264	893142	168	901642	432	098358	26
35	794942	264	893041	168	901901	432	098099	25
36	795101	264	892940	168	902160	432	097840	24
37	795259	264	892839	168	902419	432	097581	23
38	795417	263	892739	168	902679	432	097321	22
39	795575	263	892638	168	902938	432	097062	21
40	795733	263	892536	168	903197	431	096803	20
41	9·795891	263	9·892435	169	9·903455	431	10·096545	19
42	796049	263	892334	169	903714	431	096286	18
43	796206	263	892233	169	903973	431	096027	17
44	796364	262	892132	169	904232	431	095768	16
45	796521	262	892030	169	904491	431	095509	15
46	796679	262	891929	169	904750	431	095250	14
47	796836	262	891827	169	905008	431	094992	13
48	796993	262	891726	169	905267	431	094733	12
49	797150	261	891624	169	905526	431	094474	11
50	797307	261	891523	170	905784	431	094216	10
51	9·797464	261	9·891421	170	9·906043	431	10·093957	9
52	797621	261	891319	170	906302	431	093698	8
53	797777	261	891217	170	906560	431	093440	7
54	797934	261	891115	170	906819	431	093181	6
55	798091	261	891013	170	907077	431	092923	5
56	798247	261	890911	170	907336	431	092664	4
57	798403	260	890809	170	907594	431	092406	3
58	798560	260	890707	170	907852	431	092148	2
59	798716	260	890605	170	908111	430	091889	1
60	798872	260	890503	170	908369	430	091631	0
	Cosine		Sine		Cotang.		Tang.	M.

M.	Sine	D.	Cosine	D.	Tang.	D.	Cotang.	
0	9·798872	260	9·890503	170	0·908369	430	10·091631	60
1	799028	260	890400	171	908628	430	091372	59
2	799184	260	890298	171	908886	430	091114	58
3	799339	259	890195	171	909144	430	090856	57
4	799495	259	890093	171	909402	430	090598	56
5	799651	259	889990	171	909660	430	090340	55
6	799806	259	889888	171	909918	430	090082	54
7	799962	259	889785	171	910177	430	089823	53
8	800117	259	889683	171	910435	430	089565	52
9	800272	259	889579	171	910693	430	089307	51
10	800427	258	889477	171	910951	430	089049	50
11	9·800582	258	9·889374	172	0·911209	430	10·088791	49
12	800737	258	889271	172	911467	430	088533	48
13	800892	258	889168	172	911724	430	088276	47
14	801047	258	889064	172	911982	430	088018	46
15	801201	258	888961	172	912240	430	087760	45
16	801356	257	888858	172	912498	430	087502	44
17	801511	257	888755	172	912756	430	087244	43
18	801665	257	888651	172	913014	429	086986	42
19	801819	257	888548	172	913271	429	086729	41
20	801973	257	888444	173	913529	429	086471	40
21	9·802128	257	9·888341	173	0·913787	429	10·086213	39
22	802282	256	888237	173	914044	429	085956	38
23	802436	256	888134	173	914302	429	085698	37
24	802589	256	888030	173	914560	429	085440	36
25	802743	256	887926	173	914817	429	085183	35
26	802897	256	887822	173	915075	429	084925	34
27	803050	256	887718	173	915332	429	084668	33
28	803204	256	887614	173	915590	429	084410	32
29	803357	255	887510	173	915847	429	084153	31
30	803511	255	887406	174	916104	429	083896	30
31	9·803664	255	9·887302	174	9·916362	429	10·083638	29
32	803817	255	887198	174	916619	429	083381	28
33	803970	255	887093	174	916877	429	083123	27
34	804123	255	886989	174	917134	429	082866	26
35	804276	254	886885	174	917391	429	082609	25
36	804428	254	886780	174	917648	429	082352	24
37	804581	254	886676	174	917905	429	082095	23
38	804734	254	886571	174	918163	428	081837	22
39	804886	254	886466	174	918420	428	081580	21
40	805039	254	886362	175	918677	428	081323	20
41	9·805191	254	9·886257	175	0·918934	428	10·081066	19
42	805343	253	886152	175	919191	428	080809	18
43	805495	253	886047	175	919448	428	080552	17
44	805647	253	885942	175	919705	428	080295	16
45	805799	253	885837	175	919962	428	080038	15
46	805951	253	885732	175	920219	428	079781	14
47	806103	253	885627	175	920476	428	079524	13
48	806254	253	885522	175	920733	428	079267	12
49	806406	252	885416	175	920990	428	079010	11
50	806557	252	885311	176	921247	428	078753	10
51	9·806709	252	9·885205	176	9·921503	428	10·078497	9
52	806860	252	885100	176	921760	428	078240	8
53	807011	252	884994	176	922017	428	077983	7
54	807163	252	884889	176	922274	428	077726	6
55	807314	252	884783	176	922530	428	077470	5
56	807465	251	884677	176	922787	428	077213	4
57	807615	251	884572	176	923044	428	076956	3
58	807766	251	884466	176	923300	428	076700	2
59	807917	251	884360	176	923557	427	076443	1
60	808067	251	884254	177	923813	427	076187	0
	Cosine		Sine		Cotang.		Tang.	M.

M.	Sine	D.	Cosine	D.	Tang.	D.	Cotang.	
0	9·808067	251	9·884254	177	9·923813	427	10·076187	60
1	808218	251	884148	177	924070	427	075930	59
2	808368	251	884042	177	924327	427	075673	58
3	808519	250	883936	177	924583	427	075417	57
4	808669	250	883829	177	924840	427	075160	56
5	808819	250	883723	177	925096	427	074904	55
6	808969	250	883617	177	925352	427	074648	54
7	809119	250	883510	177	925609	427	074391	53
8	809269	250	883404	177	925865	427	074135	52
9	809419	249	883297	178	926122	427	073878	51
10	809569	249	883191	178	926378	427	073022	50
11	9·809718	249	9·883084	178	9·926634	427	10·073366	49
12	809868	249	882977	178	926890	427	073110	48
13	810017	249	882871	178	927147	427	072853	47
14	810167	249	882764	178	927403	427	072597	46
15	810316	248	882657	178	927659	427	072341	45
16	810465	248	882550	178	927915	427	072085	44
17	810614	248	882443	178	928171	427	071829	43
18	810763	248	882336	179	928427	427	071573	42
19	810912	248	882229	179	928683	427	071317	41
20	811061	248	882121	179	928940	427	071060	40
21	9·811210	248	9·882014	179	9·929196	427	10·070804	39
22	811358	247	881907	179	929452	427	070548	38
23	811507	247	881799	179	929708	427	070292	37
24	811655	247	881692	179	929964	426	070036	36
25	811804	247	881584	179	930220	426	069780	35
26	811952	247	881477	179	930475	426	069525	34
27	812100	247	881369	179	930731	426	069269	33
28	812248	247	881261	180	930987	426	069013	32
29	812396	246	881153	180	931243	426	068757	31
30	812544	246	881046	180	931499	426	068501	30
31	9·812692	246	9·880938	180	9·931755	426	10·068245	29
32	812840	246	880830	180	932010	426	067990	28
33	812988	246	880722	180	932266	426	067734	27
34	813135	246	880613	180	932522	426	067478	26
35	813283	246	880505	180	932778	426	067222	25
36	813430	245	880397	180	933033	426	066967	24
37	813578	245	880289	181	933289	426	066711	23
38	813725	245	880180	181	933545	426	066455	22
39	813872	245	880072	181	933800	426	066200	21
40	814019	245	879963	181	934056	426	065944	20
41	9·814166	245	9·879855	181	9·934311	426	10·065689	19
42	814313	245	879746	181	934567	426	065433	18
43	814460	244	879637	181	934823	426	065177	17
44	814607	244	879529	181	935078	426	064922	16
45	814753	244	879420	181	935333	426	064667	15
46	814900	244	879311	181	935589	426	064411	14
47	815046	244	879202	182	935844	426	064156	13
48	815193	244	879093	182	936100	426	063900	12
49	815339	244	878984	182	936355	426	063645	11
50	815485	243	878875	182	936610	426	063390	10
51	9·815631	243	9·878766	182	9·936866	425	10·063134	9
52	815778	243	878656	182	937121	425	062879	8
53	815924	243	878547	182	937376	425	062624	7
54	816069	243	878438	182	937632	425	062368	6
55	816215	243	878328	182	937887	425	062113	5
56	816361	243	878219	183	938142	425	061858	4
57	816507	242	878109	183	938398	425	061602	3
58	816652	242	877999	183	938653	425	061347	2
59	816798	242	877890	183	938908	425	061092	1
60	816943	242	877780	183	939163	425	060837	0
	Cosine		Sine		Cotang.		Tang.	M.

49 Degrees.

M.	Sine	D.	Cosine	D.	Tang.	D.	Cotang.	
0	9·816943	242	9·877780	183	9·939163	425	10·060837	60
1	817069	242	877670	183	939418	425	060582	59
2	817233	242	877560	183	939673	425	060327	58
3	817379	242	877450	183	939928	425	060072	57
4	817524	241	877340	183	940183	425	059817	56
5	817668	241	877230	184	940438	425	059562	55
6	817813	241	877120	184	940694	425	059306	54
7	817958	241	877010	184	940949	425	059051	53
8	818103	241	876899	184	941204	425	058796	52
9	818247	241	876789	184	941458	425	058542	51
10	818392	241	876678	184	941714	425	058286	50
11	9·818536	240	9·876568	184	9·941968	425	10·058032	49
12	818681	240	876457	184	942223	425	057777	48
13	818825	240	876347	184	942478	425	057522	47
14	818969	240	876236	185	942733	425	057267	46
15	819113	240	876125	185	942988	425	057012	45
16	819257	240	876014	185	943243	425	056757	44
17	819401	240	875904	185	943498	425	056502	43
18	819545	239	875793	185	943752	425	056248	42
19	819689	239	875682	185	944007	425	055993	41
20	819832	239	875571	185	944262	425	055738	40
21	9·819976	239	9·875459	185	9·944517	425	10·055483	39
22	820120	239	875348	185	944771	424	055229	38
23	820263	239	875237	185	945026	424	054974	37
24	820406	239	875126	186	945281	424	054719	36
25	820550	238	875014	186	945535	424	054465	35
26	820693	238	874903	186	945790	424	054210	34
27	820836	238	874791	186	946045	424	053955	33
28	820979	238	874680	186	946299	424	053701	32
29	821122	238	874568	186	946554	424	053446	31
30	821265	238	874456	186	946808	424	053192	30
31	9·821407	238	9·874344	186	9·947063	424	10·052937	29
32	821550	238	874232	187	947318	424	052682	28
33	821693	237	874121	187	947572	424	052428	27
34	821835	237	874009	187	947826	424	052174	26
35	821977	237	873896	187	948081	424	051919	25
36	822120	237	873784	187	948336	424	051664	24
37	822262	237	873672	187	948590	424	051410	23
38	822404	237	873560	187	948844	424	051156	22
39	822546	237	873448	187	949099	424	050901	21
40	822688	236	873335	187	949353	424	050647	20
41	9·822830	236	9·873223	187	9·949607	424	10·050393	19
42	822972	236	873110	188	949862	424	050138	18
43	823114	236	872998	188	950116	424	049884	17
44	823255	236	872885	188	950370	424	049630	16
45	823397	236	872772	188	950625	424	049375	15
46	823539	236	872659	188	950879	424	049121	14
47	823680	235	872547	188	951133	424	048867	13
48	823821	235	872434	188	951388	424	048612	12
49	823963	235	872321	188	951642	424	048358	11
50	824104	235	872208	188	951896	424	048104	10
51	9·824245	235	9·872095	189	9·952150	424	10·047850	9
52	824386	235	871981	189	952405	424	047595	8
53	824527	235	871868	189	952659	424	047341	7
54	824668	234	871755	189	952913	424	047087	6
55	824808	234	871641	189	953167	423	046833	5
56	824949	234	871528	189	953421	423	046579	4
57	825090	234	871414	189	953675	423	046325	3
58	825230	234	871301	189	953929	423	046071	2
59	825371	234	871187	189	954183	423	045817	1
60	825511	234	871073	190	954437	423	045563	0
	Cosine		Sine		Cotang.		Tang.	M.

M.	Sine	D.	Cosine	D.	Tang.	D.	Cotang.	
0	9·825511	234	9·871073	190	9·954437	423	10·045563	60
1	825651	233	870960	190	954691	423	045309	59
2	825791	233	870846	190	954945	423	045055	58
3	825931	233	870732	190	955200	423	044800	57
4	826071	233	870618	190	955454	423	044546	56
5	826211	233	870504	190	955707	423	044293	55
6	826351	233	870390	190	955961	423	044039	54
7	826491	233	870276	190	956215	423	043785	53
8	826631	233	870161	190	956469	423	043531	52
9	826770	239	870047	191	956723	423	043277	51
10	826910	239	869933	191	956977	423	043023	50
11	9·827049	239	9·869818	191	9·957231	423	10·042769	49
12	827189	239	869704	191	957485	423	042515	48
13	827328	239	869589	191	957739	423	042261	47
14	827467	239	869474	191	957993	423	042007	46
15	827606	239	869360	191	958246	423	041754	45
16	827745	239	869245	191	958500	423	041500	44
17	827884	231	869130	191	958754	423	041246	43
18	828023	231	869015	192	959008	423	040992	42
19	828162	231	868900	192	959262	423	040738	41
20	828301	231	868785	192	959516	423	040484	40
21	9·828439	231	9·868670	192	9·959769	423	10·040231	39
22	828578	231	868555	192	960023	423	039977	38
23	828716	231	868440	192	960277	423	039723	37
24	828855	230	868324	192	960531	423	039469	36
25	828993	230	868209	192	960784	423	039216	35
26	829131	230	868093	192	961038	423	038962	34
27	829269	230	867978	193	961291	423	038709	33
28	829407	230	867862	193	961545	423	038455	32
29	829545	230	867747	193	961799	423	038201	31
30	829683	230	867631	193	962052	423	037948	30
31	9·829821	229	9·867515	193	9·962306	423	10·037694	29
32	829959	229	867399	193	962560	423	037440	28
33	830097	229	867283	193	962813	423	037187	27
34	830234	229	867167	193	963067	423	036933	26
35	830372	229	867051	193	963320	423	036680	25
36	830509	229	866935	194	963574	423	036426	24
37	830646	229	866819	194	963827	423	036173	23
38	830784	229	866703	194	964081	423	035919	22
39	830921	228	866586	194	964335	423	035665	21
40	831058	228	866470	194	964588	422	035412	20
41	9·831195	228	9·866353	194	9·964842	422	10·035158	19
42	831332	228	866237	194	965095	422	034905	18
43	831469	228	866120	194	965349	422	034651	17
44	831606	228	866004	195	965602	422	034398	16
45	831742	228	865887	195	965855	422	034145	15
46	831879	228	865770	195	966109	422	033891	14
47	832015	227	865653	195	966362	422	033638	13
48	832152	227	865536	195	966616	422	033384	12
49	832288	227	865419	195	966869	422	033131	11
50	832425	227	865302	195	967123	422	032877	10
51	9·832561	227	9·865185	195	9·967376	422	10·032624	9
52	832697	227	865068	195	967629	422	032371	8
53	832833	227	864950	195	967883	422	032117	7
54	832969	226	864833	196	968136	422	031864	6
55	833105	226	864716	196	968389	422	031611	5
56	833241	226	864598	196	968643	422	031357	4
57	833377	226	864481	196	968896	422	031104	3
58	833512	226	864363	196	969140	422	030851	2
59	833648	226	864245	196	969403	422	030597	1
60	833783	226	864127	196	969656	422	030344	0
	Cosine		Sine		Cotang.		Tang.	M.

47 Degrees.

M.	Sine	D.	Cosine	D.	Tang.	D.	Cotang.	
0	9·833783	226	9·864127	196	9·969656	422	10·030344	60
1	833919	225	864010	196	969909	422	030091	59
2	834054	225	863902	197	970162	422	029838	58
3	834189	225	863774	197	970416	422	029584	57
4	834325	225	863656	197	970669	422	029331	56
5	834460	225	863538	197	970922	422	029078	55
6	834595	225	863419	197	971175	422	028825	54
7	834730	225	863301	197	971429	422	028571	53
8	834865	225	863183	197	971682	422	028318	52
9	834999	224	863064	197	971935	422	028065	51
10	835134	224	862946	198	972188	422	027812	50
11	9·835269	224	9·862827	198	9·972441	422	10·027559	49
12	835403	224	862709	198	972694	422	027306	48
13	835538	224	862590	198	972948	422	027052	47
14	835672	224	862471	198	973201	422	026799	46
15	835807	224	862353	198	973454	422	026546	45
16	835941	224	862234	198	973707	422	026293	44
17	836075	223	862115	198	973960	422	026040	43
18	836209	223	861996	198	974213	422	025787	42
19	836343	223	861877	198	974466	422	025534	41
20	836477	223	861758	199	974719	422	025281	40
21	9·836611	223	9·861638	199	9·974973	423	10·025027	39
22	836745	223	861519	199	975226	422	024774	38
23	836878	223	861400	199	975479	422	024521	37
24	837012	222	861280	199	975732	422	024268	36
25	837146	222	861161	199	975985	422	024015	35
26	837279	222	861041	199	976238	422	023762	34
27	837412	222	860922	199	976491	422	023509	33
28	837546	222	860802	199	976744	423	023256	32
29	837679	222	860682	200	976997	422	023003	31
30	837812	222	860562	200	977250	422	022750	30
31	9·837945	222	9·860442	200	9·977503	422	10·022497	29
32	838078	221	860322	200	977756	422	022244	28
33	838211	221	860202	200	978009	422	021991	27
34	838344	221	860082	200	978262	422	021738	26
35	838477	221	859962	200	978515	422	021485	25
36	838610	221	859842	200	978768	422	021232	24
37	838742	221	859721	201	979021	422	020979	23
38	838875	221	859601	201	979274	422	020726	22
39	839007	221	859480	201	979527	422	020473	21
40	839140	220	859360	201	979780	422	020220	20
41	9·839272	220	9·859239	201	9·980033	422	10·019967	19
42	839404	220	859118	201	980286	422	019714	18
43	839536	220	858998	201	980538	422	019462	17
44	839668	220	858877	201	980791	421	019209	16
45	839800	220	858756	202	981044	421	018956	15
46	839932	220	858635	202	981297	421	018703	14
47	840064	219	858514	202	981550	421	018450	13
48	840196	219	858393	202	981803	421	018197	12
49	840328	219	858272	202	982056	421	017944	11
50	840459	219	858151	202	982309	421	017691	10
51	9·840591	219	9·858029	202	9·982562	421	10·017438	9
52	840722	219	857908	202	982814	421	017186	8
53	840854	219	857786	202	983067	421	016933	7
54	840985	219	857665	203	983320	421	016680	6
55	841116	218	857543	203	983573	421	016427	5
56	841247	218	857422	203	983826	421	016174	4
57	841378	218	857300	203	984079	421	015921	3
58	841509	218	857178	203	984331	421	015669	2
59	841640	218	857056	203	984584	421	015416	1
60	841771	218	856934	203	984837	421	015163	0
	Cosine		Sine		Cotang.		Tang.	M.

46 Degrees.

M.	Sine	D	Cosine	D.	Tang.	D.	Cotang.	
0	9·841771	218	9·856934	203	9·984837	421	10·015163	60
1	841902	218	856812	203	985090	421	014910	59
2	842033	218	856690	204	985343	421	014657	58
3	842163	217	856568	204	985596	421	014404	57
4	842294	217	856446	204	985848	421	014152	56
5	842424	217	856323	204	986101	421	013899	55
6	842555	217	856201	204	986354	421	013646	54
7	842685	217	856078	204	986607	421	013393	53
8	842815	217	855956	204	986860	421	013140	52
9	842946	217	855833	204	987112	421	012888	51
10	843076	217	855711	205	987365	421	012635	50
11	9·843206	216	9·855588	205	9·987618	421	10·012382	49
12	843336	216	855465	205	987871	421	012129	48
13	843466	216	855342	205	988123	421	011877	47
14	843595	216	855219	205	988376	421	011624	46
15	843725	216	855096	205	988629	421	011371	45
16	843855	216	854973	205	988882	421	011118	44
17	843984	216	854850	205	989134	421	010866	43
18	844114	215	854727	206	989387	421	010613	42
19	844243	215	854603	206	989640	421	010360	41
20	844372	215	854480	206	989893	421	010107	40
21	9·844502	215	9·854356	206	9·990145	421	10·009855	39
22	844631	215	854233	206	990398	421	009602	38
23	844760	215	854109	206	990651	421	009349	37
24	844889	215	853986	206	990903	421	009097	36
25	845018	215	853862	206	991156	421	008844	35
26	845147	215	853738	206	991409	421	008591	34
27	845276	214	853614	207	991662	421	008338	33
28	845405	214	853490	207	991914	421	008086	32
29	845533	214	853366	207	992167	421	007833	31
30	845662	214	853242	207	992420	421	007580	30
31	9·845790	214	9·853118	207	9·992672	421	10·007328	29
32	845919	214	852994	207	992925	421	007075	28
33	846047	214	852869	207	993178	421	006822	27
34	846175	214	852745	207	993430	421	006570	26
35	846304	214	852620	207	993683	421	006317	25
36	846432	213	852496	208	993936	421	006064	24
37	846560	213	852371	208	994189	421	005811	23
38	846688	213	852247	208	994441	421	005559	22
39	846816	213	852122	208	994694	421	005306	21
40	846944	213	851997	208	994947	421	005053	20
41	9·847071	213	9·851872	208	9·995199	421	10·004801	19
42	847199	213	851747	208	995452	421	004548	18
43	847327	213	851622	208	995705	421	004295	17
44	847454	212	851497	209	995957	421	004043	16
45	847582	212	851372	209	996210	421	003790	15
46	847709	212	851246	209	996463	421	003537	14
47	847836	212	851121	209	996715	421	003285	13
48	847964	212	850996	209	996968	421	003032	12
49	848091	212	850870	209	997221	421	002779	11
50	848218	212	850745	209	997473	421	002527	10
51	9·848345	212	9·850619	209	9·997726	421	10·002274	9
52	848472	211	850493	210	997979	421	002021	8
53	848599	211	850368	210	998231	421	001769	7
54	848726	211	850242	210	998484	421	001516	6
55	848852	211	850116	210	998737	421	001263	5
56	848979	211	849990	210	998989	421	001011	4
57	849106	211	849864	210	999242	421	000758	3
58	849232	211	849738	210	999495	421	000505	2
59	849359	211	849611	210	999748	421	000253	1
60	849485	211	849485	210	10·000000	421	000000	0
	Cosine		Sine		Cotang.		Tang.	M.

45 Degrees.